Applications of Optical Measurement Techniques in Aerospace Research

snova
www.novapublishers.com
New York

Aircraft and Avionics Research and Technology Series

Aircraft: Its Development in War and Peace and Its Commercial Future
Evan John David (Editor)
ISBN: 979-8-89530-588-1
eBook ISBN: 979-8-89530-804-2

Safety and Design of the Boeing 737 Max: Were the Crashes Avoidable?
Malthe E. Ravn (Editor)
ISBN: 978-1-53617-902-6
eBook ISBN: 978-1-53617-903-3

The Status of the Boeing 737 Max and Flight Control System Review
John I. Corneau (Editor)
ISBN: 978-1-53617-839-5
eBook ISBN: 978-1-53617-840-1

Aviation and the Airline Industry: International Perspectives on Policies and Practices
Leo Kounis (Editor)
ISBN: 978-1-53616-937-9
eBook ISBN: 978-1-53616-938-6

More information about this series can be found at

Octavian Melina
Editor

Applications of Optical Measurement Techniques in Aerospace Research

nova
www.novapublishers.com
New York

Library of Congress Cataloging-in-Publication Data

ISBN: 979-8-90134-275-6 (Softcover)
ISBN: 979-8-90134-294-7 (eBook)

Published by Nova Science Publishers, Inc. † New York

Contents

Preface

Optical measurement technologies play a critical role in advancing aerospace research by enabling precise monitoring of structures and materials under dynamic conditions. This collection highlights recent applications of fiber-optic and optical sensing techniques. Chapter 1 examines Doppler fiber-optic sensors for structural velocity estimation, providing high-fidelity dynamic data. Complementing this, fiber-optic sensors are applied to calibrate loads on a mock truss-braced wing, demonstrating their value for structural health and aerodynamic analysis in chapter 2. Chapter 3 extends optical diagnostics to spacecraft materials, analyzing the HORIS windowpane under flight conditions. Together, these studies illustrate the versatility of optical sensing in aerospace, bridging experimental methods with practical applications in structural evaluation, flight testing, and materials performance.

Chapter 1

The Examination of Doppler Fiber Optic Sensing for Structural Velocity Estimations*

Francisco Peña
Benjamin C. Park
and Alexander Chin

NASA STI Program Report Series

The NASA STI Program collects, organizes, provides for archiving, and disseminates NASA's STI. The NASA STI program provides access to the NTRS Registered and its public interface, the NASA Technical Reports Server, thus providing one of the largest collections of aeronautical and space science STI in the world. Results are published in both non- NASA channels and by NASA in the NASA STI Report Series, which includes the following report types:

- TECHNICAL PUBLICATION. Reports of completed research or a major significant phase of research that present the results of NASA Programs and include extensive data or theoretical analysis. Includes compilations of significant scientific and technical data and information deemed to be of continuing reference value. NASA counter-part of peer- reviewed formal professional papers but has less stringent limitations on manuscript length and extent of graphic presentations.

---

* This is an edited, reformatted and augmented version of NASA Publication No. NASA/TM–20210025579, dated June 2022.

In: Applications of Optical Measurement Techniques …
Editor: Octavian Melina
ISBN: 979-8-90134-275-6

- TECHNICAL MEMORANDUM.
  Scientific and technical findings that are preliminary or of specialized interest, e.g., quick release reports, working papers, and bibliographies that contain minimal annotation. Does not contain extensive analysis.
- CONTRACTOR REPORT. Scientific and technical findings by NASA-sponsored contractors and grantees.
- CONFERENCE PUBLICATION.
  Collected papers from scientific and technical conferences, symposia, seminars, or other meetings sponsored or co-sponsored by NASA.
- SPECIAL PUBLICATION. Scientific, technical, or historical information from NASA programs, projects, and missions, often concerned with subjects having substantial public interest.
- TECHNICAL TRANSLATION.
  English-language translations of foreign scientific and technical material pertinent to NASA's mission.

Specialized services also include organizing and publishing research results, distributing specialized research announcements and feeds, providing information desk and personal search support, and enabling data exchange services.

For more information about the NASA STI program, see the following:

- Access the NASA STI program home page at http://www.sti.nasa.gov
- Help desk contact information:
- https://www.sti.nasa.gov/sti-contact-form/
- and select the "General" help request type.

Abstract

Using a Fiber Optic Sensing System interrogator and a Doppler Fiber Optic Sensing System analysis algorithm, estimates of the distributed structural velocity information can be calculated throughout the entire span of a structure where distributed fiber Bragg gratings are installed. The Doppler Fiber Optic Sensing System algorithm can provide full state feedback control, based on direct structural measurements, which could improve controller robustness and performance. These improvements also enable implementation for control laws with applications such as controlling flexible vehicles. This research effort covers the algorithm developmental work and the ground vibration test using the X-56A Multi-Utility Technology Testbed right wing bare structure to examine the feasibility of utilizing the Doppler Fiber Optic Sensing System for future control applications.

Nomenclature

Accel	accelerometer mode
FBG	fiber Bragg grating
FFT	fast Fourier transform
FOSS	Fiber Optic Sensing System
GVT	ground vibration test
OFDR	optical frequency domain reflectometry
SVD	Singular Value Decomposition
Vel	velocity

Introduction

Aircraft flight control systems typically use feedback control sensors that measure parameters such as vehicle attitude, pressure, and acceleration (ref. 1). These sensors work well for controlling the flight dynamics of a rigid vehicle. Increasingly, flexible vehicles require consideration of the vehicle structural dynamics to ensure adverse control system interaction is avoided, and if necessary, the vehicle structural dynamics are actively controlled as well. Multi-purpose control systems would control both the vehicle flight dynamics and structural dynamics [1].

Real-time linear structural velocity measurement capabilities are sparse. Sensors such as gyroscopes can measure angular rates, but not linear rates. Laser vibrometers can measure structural velocity, provided there is appropriate field of view between the laser and reference point. Velocity can be calculated from differentiating position changes or integrating accelerometer measurements, but the process can amplify noise and introduce time delays. It is uncommon, therefore, to use direct structural velocity measurements for modal feedback control because estimations are either insufficient, or existing measurements systems are not reliable for flight applications.

Fiber Optic Sensing System

The Fiber Optic Sensing System (FOSS) that was developed at the National Aeronautics and Space Administration (NASA) Armstrong Flight Research Center (AFRC) (Edwards, California) was utilized in conjunction with optical frequency domain reflectometry (OFDR) [2] methods to obtain the data presented in this report. The FOSS can be used to interrogate optical fiber with a series up to 1,000 (for typical applications) distributed fiber Bragg gratings (FBGs) to provide FBG response data. By utilizing the OFDR methods with the FOSS, the distributed FBG data can be used to determine both the static vehicle deflection [3] and estimates of the structural deflection velocity through observation of optical Doppler effects. Doppler frequency shifts are observed when a reflective surface is moving relative to the static observer.

While technology exists to obtain structural deflection velocity data in a ground test setting, e.g., using laser vibrometers, a sensing system of this type is impractical in flight because of line of sight and visibility restrictions (among other variables that could contribute to contamination of this data). The ability of the Doppler FOSS to estimate velocity and displacements in real time may enable full structural state feedback for active structural controllers; thus, allowing suppression of unsteady structural modes.

Doppler Fiber Optic Sensing System Algorithm

The Doppler FOSS algorithm is used to process the data that is produced by raw optical light that is reflected by the distributed FBGs, called the interferogram, and is observed by the OFDR FOSS interrogator, as shown in figures 1. An example of the interferogram is shown in figure 1(a); the horizontal axis is the wavelength of light, and the vertical axis is the amplitude of the light. Using the OFDR FOSS method, each FBG has its own carrier frequency, which under static conditions, is proportional to its unique distance down the length of a fiber-optic cable. After using a fast discrete Fourier transform (DFT) of the interferogram, each FBG appears as a top hat signal in the frequency domain. While experiencing dynamic strain, the carrier frequency of each FBG also experiences a frequency shift, referred to as a Doppler shift, that is proportional to the integrated strain along the length of the sensing fiber line. A schematic of the Doppler FOSS algorithm is shown in Figure 2.

A comparison between the static carrier frequency and the carrier frequency shift as a result of Doppler effects is shown in figure 1(b); the X-axis represents the fast Fourier transform (FFT) frequency count of the interferogram, and the Y-axis represents the absolute value of the FFT of the interferogram. In figure 1(b), the black curve represents the static carrier frequency, and the red curve represents the Doppler-shifted carrier frequency. The difference in the shifts in frequency are proportional to the integrated strain along the length of the fiber. Differentiating the Doppler shift frequency component with respect to the span of the structure provides a signal that is proportional to strain rate (strain per unit time). Once strain rate signals and structural geometry parameters for the top and bottom surfaces are obtained, it is then possible to implement beam curvature equations [3] for estimation of the structural velocity profile along the span. The frequency counts are scaled by 1/1,000,000 to obtain units (analogous to microstrain) that are compatible with the strain-based shape-sensing algorithms. An example of the spanwise structural velocity profile is shown in figure 1(c); the X-axis represents the FBG number along the span (analogous to the length), and the Y-axis represents the approximate normalized-vertical velocity profile of the structure at each FBG location. The time-history trace of the wing tip velocity, based on measurement frames, is shown in figure 1(d); the X-axis represents the frame number that is analogous to time, and the Y-axis represents the time trace of the distal tip velocity history.

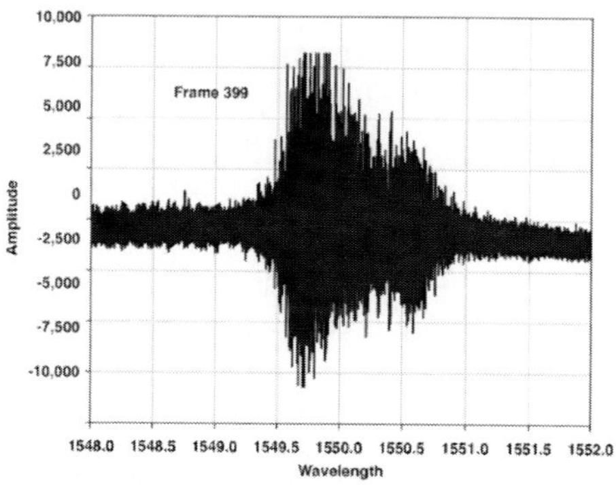

(a) Interferogram

Figure 1. (Continued).

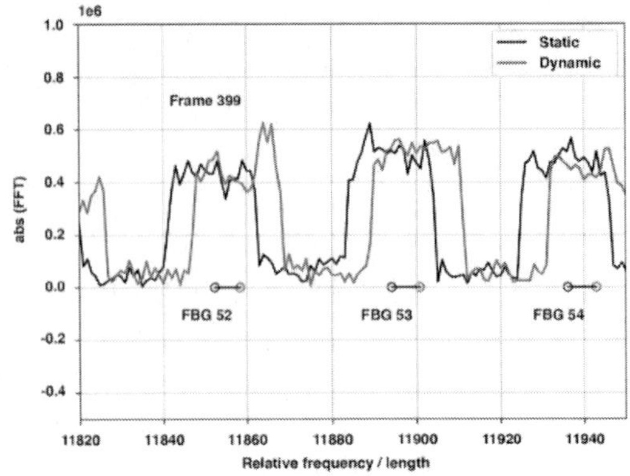

(b) Frequency domain; top hats shift as a result of Doppler effects

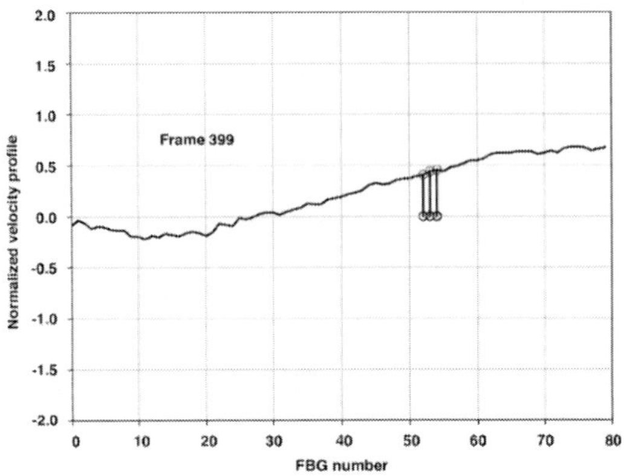

(c) Velocity shape profile of the vibrating structure

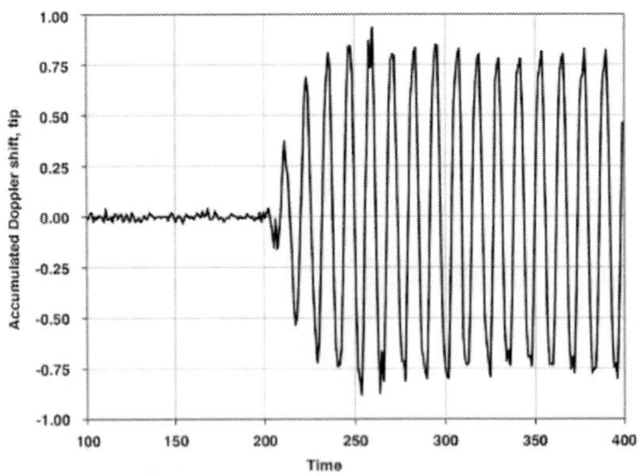

(d) Structure tip velocity time trace (represented in frames)

220000

Figure 1. Example algorithm process during sine dwell shaker excitation.

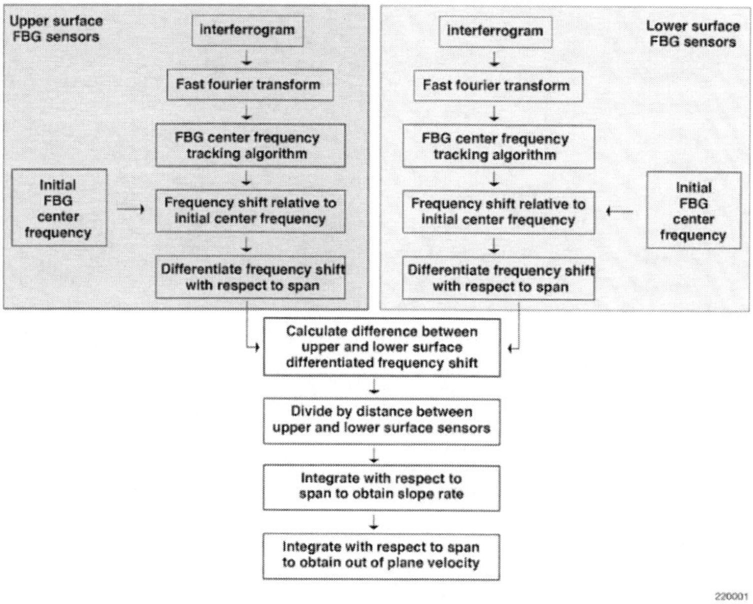

Figure 2. Doppler Fiber Optic Sensing System algorithm diagram.

Ground Vibration Test

This research effort included a ground vibration test (GVT) that was conducted on the X-56A Multi-Utility Technology Testbed (Lockheed Martin) (Bethesda, Maryland) right wing bare structure.

The right wing was unpowered and had no control surfaces or control surface actuators, which contribute to the overall mass and stiffness of the wing. The wing was cantilevered off a strongback test fixture. A shaker was connected to the wing tip leading edge and was used to excite the structure. The X-56A Multi-Utility Technology Testbed right wing was instrumented with distributed FBGs along the forward spar on the top and bottom and was connected to the FOSS interrogator to obtain raw interferogram signals at a rate of approximately 50.0 Hz. For this experiment, there were 203-distributed FBGs along the 160-inch forward spar span. An image of the X-56A Multi-Utility Technology Testbed and its right wing instrumented with FBGs is shown on the right in figure 3. An image of the test setup is shown in figure 4.

Figure 3. (left) X-56A Multi-Utility Technology Testbed; and (right) the right wing instrumented with fiber Bragg gratings and accelerometers on the top surface, during ground vibration test.

The interferograms were processed by the Doppler FOSS algorithm posttest. The wing was instrumented with distributed triaxial accelerometers and rate gyros, and a laser Doppler vibrometer system was utilized during the testing sequence. The wing tip accelerometer was integrated to provide velocity time-history data for the Doppler FOSS algorithm to compare against. Rate gyros were compared against the intermediate shape calculations of the Doppler FOSS algorithm which provided slope rates.

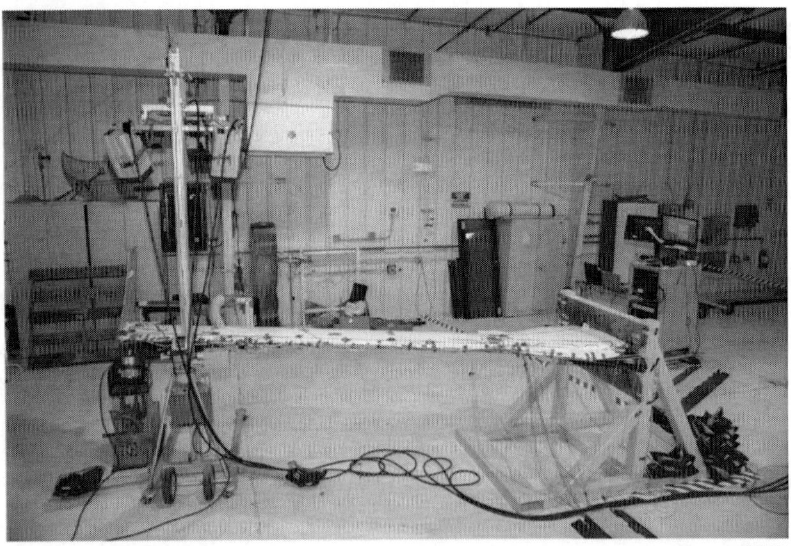

220003

Figure 4. Doppler Fiber Optic Sensing System X-56A Multi-Utility Technology Testbed right wing bare structure ground vibration test setup.

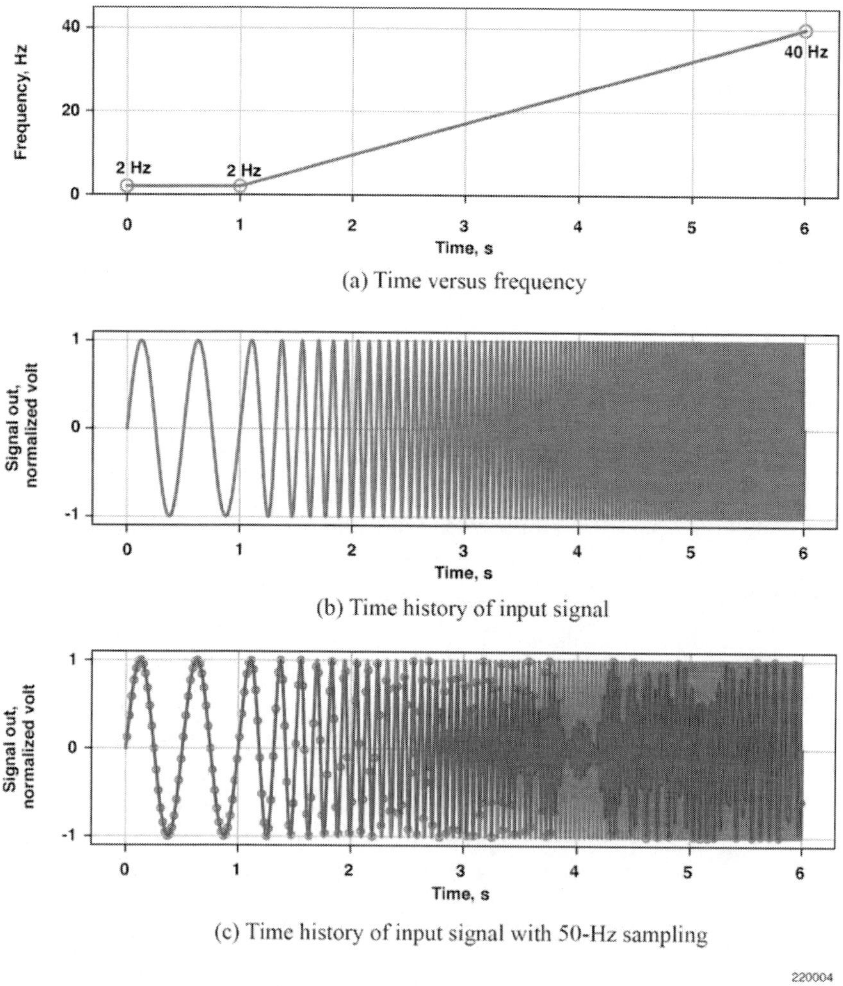

Figure 5. Sine sweep input signal.

The Doppler vibrometer system was used to estimate the mode shapes but was not used for time-history data.

The input voltage to the GVT shaker was a short dwell of 2.0 Hz for one second, followed by a sine sweep from 2.0 to 40.0 Hz over the following five seconds. An image of the sine sweep input signal and frequency content are shown in figure 5. Figure 5(a) shows the time history of the input signal; figures5(b) and 5(c) show how frequency changes with respect to time. Because of the limitations of the laser technology within the FOSS interrogator, the acquisition rate is limited to approximately 50.0 Hz. The relatively low sample rate of the FOSS caused aliasing at the higher frequency portion of the input sine sweep. An example of the aliasing caused by incompatible sample rates can be seen in the resampled input sine sweep signal (sampled at 50.0 Hz), shown in figure 5(c). The peak amplitudes of an oscillating signal were not consistently captured when the sensor did not have an adequate sampling rate. Though aliasing in the time-history domain was anticipated, there was still spatial data that could be extracted from the aliased signal in the form of mode shapes.

Test Results

The interferograms collected during the GVT were processed using the Doppler FOSS algorithm.

Concerning span, the strain rates were integrated to obtain distributed angular slope rates and were scaled linearly by a calibration coefficient for comparison against rate gyro readings. The time histories of wing tip slope rates estimated by the Doppler FOSS algorithm and independent rate gyro measurements are shown in figure 6. The wing tip rate gyro used in this test had a dynamic range of 50 degrees per second, and for this test series, the rate gyro data was assumed to be the truth source in the comparison.

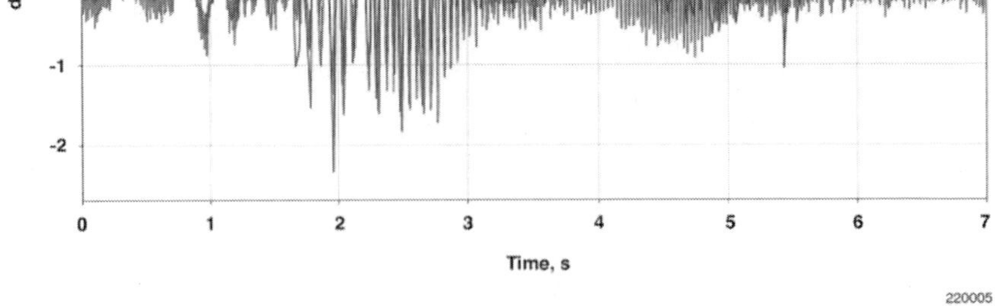

Figure 6. Wing tip slope rate comparisons between Doppler Fiber Optic Sensing System and rate gyro.

Concerning span, the angular slope rates from the FOSS measurements were then integrated to obtain out-of-plane velocities. The time histories of the wing tip velocity estimated by the Doppler FOSS algorithm, and velocity estimates derived from wing tip accelerometers are shown in figure 7. The Doppler FOSS algorithm and implementation assumes that the Doppler shifts observed were caused by vertical deformations. The FBG sensors, however, respond to axial strain along the fiber direction, regardless of whether they were derived from vertical deflection or lateral deflection; therefore, the Doppler FOSS algorithm provided velocity outputs that were sensitive to both vertical deflection and forward-aft motion. The vertical velocities were obtained by comparing the difference in strain rate on the top and bottom surfaces [3]; however, this process was not sufficient for separating the vertical and lateral velocities, as shown in figure 7. The Doppler FOSS algorithm was responsive to both vertical and forward-aft wing motions, resulting in erroneous high-velocity estimates during forward-aft dominant modes. Further investigation is required to explore methods to decouple the vertical velocity profile from the lateral velocity profile using the Doppler FOSS algorithm.

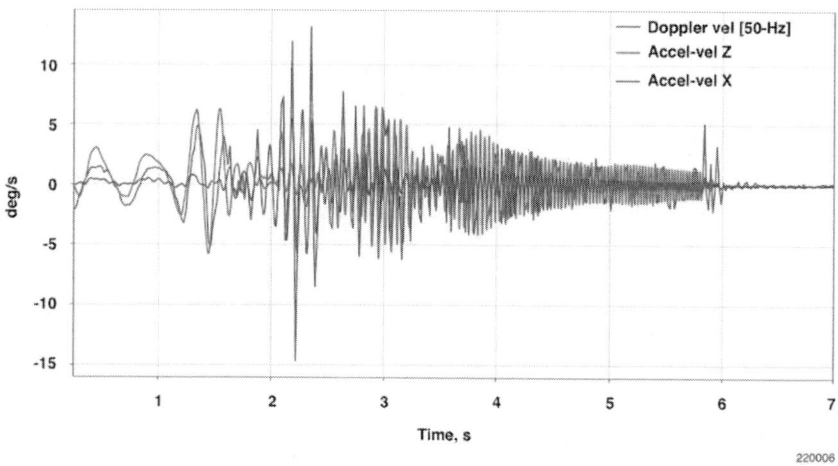

Figure 7. Wing tip velocity comparisons between Doppler Fiber Optic Sensing System and wing tip accelerometer-derived velocities in the vertical (Z) and lateral (X) directions.

For each velocity time-trace data point, as shown in figure 7, the Doppler FOSS can provide the full spanwise profile of the velocity estimates. A useful method to reduce the data set is to perform modal decomposition to extract the dominant mode shapes from the data. One technique used for modal extraction is the Singular Value Decomposition (SVD) signal processing method [4, 5].

The first three modes of the wing were determined using accelerometer measurements. The mode shapes shown in figure 8 are the first bending, first torsion, and second bending at the approximate, corresponding frequencies of 4.0, 10.0, and 12.0 Hz, respectively. Subsequent mode shapes and time traces, using the Doppler FOSS and laser vibrometers, were extracted at these frequencies for comparison. The Doppler FOSS algorithm measures in the vertical direction only, or one degree of freedom; thus, limiting the rendering of the calculated mode shapes to reflect bending measurements exclusively. Independent accelerometer measurements provided further insight into the behavior of the wing in degrees of freedom, not captured by the Doppler FOSS (i.e., torsion and forward/aft motion).

The first three velocity bending mode shapes from the Doppler FOSS, extracted using the SVD technique are shown in figure 9. The SVD technique breaks down the entire two-dimensional (2D) data array signal into dominant sensor shape profiles and time-history traces for each shape profile; rows represent each scan, and columns represent time history of each FBG. The time traces of the first three dominant SVD modes are shown in figure 10. The summation of all the decomposed time-history traces will reconstruct the original signal. The time-trace data shows that the majority of the deformation is defined by the first SVD mode shape.

(a) 1st mode (first bending), 4.0 Hz

(b) 2nd mode (first torsion), 10.0 Hz

(c) 3rd mode (second bending), 12.0 Hz

Figure 8. First three bending/torsion mode shapes of the wing.

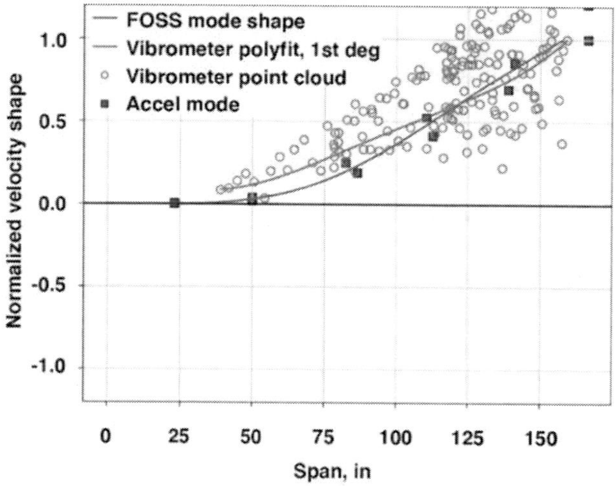

(a) Singular Value Decomposition mode shape 1 (first bending)

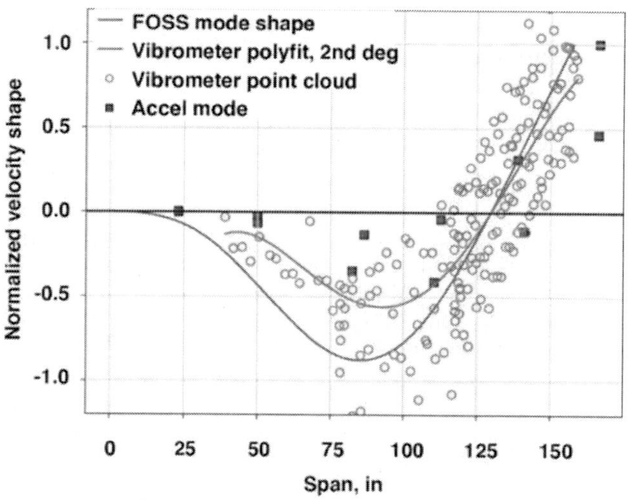

(b) Singular Value Decomposition mode shape 2 (second bending)

Figure 9. (Continued).

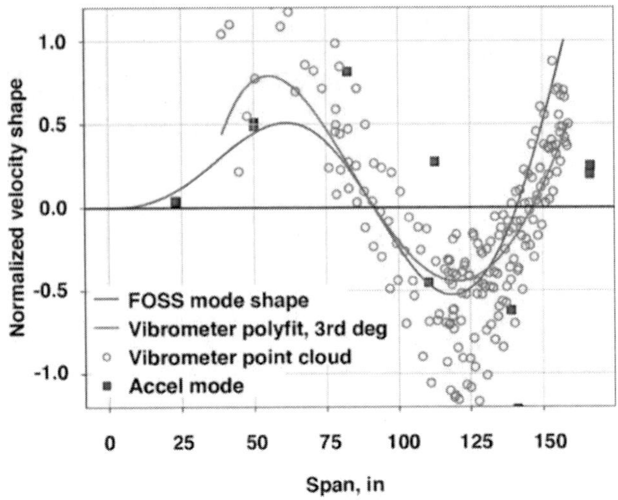

(c) Singular Value Decomposition
mode shape 3 (third bending)

220008

Figure 9. Normalized mode shapes extracted from Singular Value Decomposition of Doppler Fiber Optic Sensing System data.

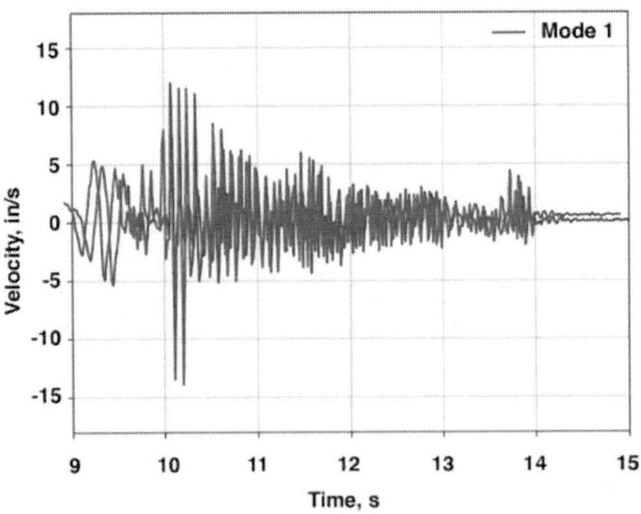

(a) Time trace Singular Value Decomposition
mode shape 1

Figure 10. (Continued).

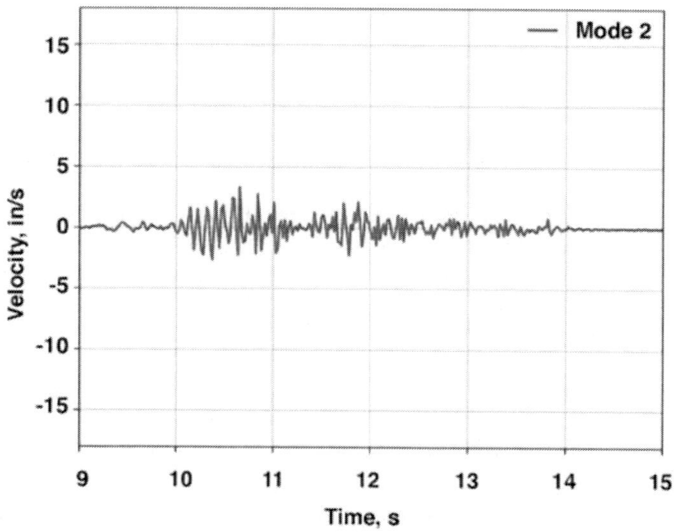

(b) Time trace Singular Value Decomposition mode shape 2

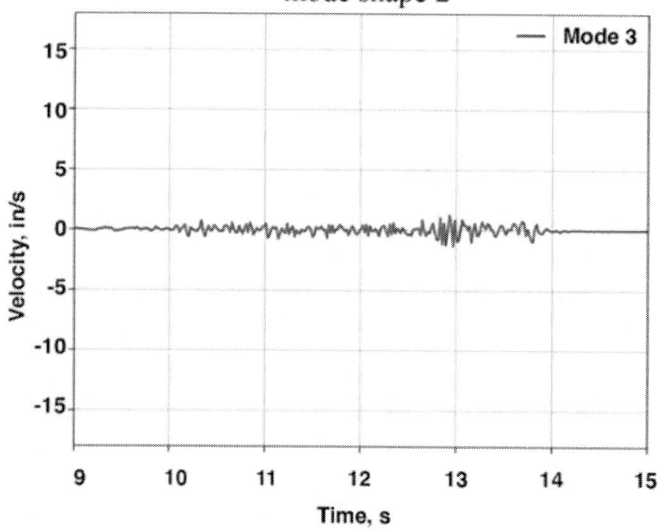

(c) Time trace Singular Value Decomposition mode shape 3

220009

Figure 10. Normalized mode shapes extracted from time-trace Singular Value Decomposition of Doppler Fiber Optic Sensing System data.

Conclusion

This research effort utilized an experimental Fiber Optic Sensing System during a ground vibration test of the X-56A Multi-Utility Technology Testbed experimental aircraft right wing bare structure. The Fiber Optic Sensing System was able to obtain interferogram scans at an approximate rate of 50.0 Hz. These interferograms were processed by the Doppler Fiber Optic Sensing System algorithm and combined with strain-based shape-sensing algorithms to provide estimates of the distributed velocity profiles along the entire wingspan. The Doppler Fiber Optic Sensing System algorithm can be integrated into the Fiber Optic Sensing System central processing unit to provide real-time span-wise velocity information to the user.

Independent sensors such as accelerometers, rate gyros, and non-contact laser Doppler vibrometer systems were utilized to provide independent measurements during the ground vibration test. Side-by-side measurement comparisons were provided for both the Doppler Fiber Optic Sensing System slope rate, utilizing rate gyros; and the Doppler Fiber Optic Sensing System tip velocity, utilizing vertical velocity derived from a wing tip mounted accelerometer. Mode shapes were provided from the Doppler Fiber Optic Sensing System velocities, and the mode shape point cloud data was provided by the laser vibrometer.

Because the Doppler Fiber Optic Sensing System technology is still under development, current state-of-the-art swept laser technology limits the acquisition rate of the Fiber Optic Sensing System to approximately 50.0 Hz. Advances in swept laser technology are required to speed up the acquisition rates and reduce the exposure time during the acquisition of each scan; thus, reducing blur and further improving system measurements. Further investigation is required to explore methods to decouple the vertical velocity from the lateral motions using the Doppler Fiber Optic Sensing System algorithm.

Because the Fiber Optic Sensing System can monitor distributed fiber Bragg grating data in real time, it can be used to determine both the static vehicle deflection and estimate the structural deflection velocity to provide real-time data to a multi-purpose flight control system; thus, reducing the structural dynamic deformations of highly flexible aircraft wings.

References

[1] Ryan, John J., John T. Bosworth, John J. Burken, and Peter M. Suh, "Current and Future Research in Active Control of Lightweight, Flexible Structures Using the X-56 Aircraft," AIAA Paper 2014-0597, 2014.

[2] Richards, W. Lance, Allen R. Parker, William L. Ko, Anthony Piazza, and Patrick Chan, "Application of Fiber Optic Instrumentation," NATO Research and Technology Organization (RTO), RTO AGARDograph 160, *Flight Test Instrumentation Series* – Volume 22, RTO-AG-160-V22, SCI-228, July 2012.

[3] Peña, Francisco, Fiber-Optic Strain-Based Deflection and Twist Sensing for a High-Aspect-Ratio Swept Wing, *NASA/TM-2020-220465, 2020.*

[4] Feeny, B. F., R. Kappagantu, "On the Physical Interpretation of Proper Orthogonal Modes in Vibrations," *Journal of Sound and Vibration*, Vol. 211, Issue 4, April 1998, pp. 607-616.

[5] Fahnline, J. B., R. L. Campbell, and S. A. Hambric, "Modal Analysis Using the Singular Value Decomposition," *Technical Report,* U.S. *Department of Energy Office of Scientific and Technical Information (OSTI), OST 836294*, No. LM-04K005, TRN: US200503%%555, February 2004.

Chapter 2

Mock Truss-Braced Wing Loads Calibration Research Utilizing Fiber-Optic Sensors*

Francisco Peña
and Benjamin Park

NASA STI Program Report Series

Since its founding, NASA has been dedicated to the advancement of aeronautics and space science. The NASA scientific and technical information (STI) program plays a key part in helping NASA maintain this important role.

The NASA STI program operates under the auspices of the Agency Chief Information Officer. It collects, organizes, provides for archiving, and disseminates NASA's STI. The NASA STI program provides access to the NTRS Registered and its public interface, the NASA Technical Reports Server, thus providing one of the largest collections of aeronautical and space science STI in the world. Results are published in both non-NASA channels and by NASA in the NASA STI Report Series, which includes the following report types:

TECHNICAL PUBLICATION. Reports of completed research or a major significant phase of research that present the results of NASA Programs and include extensive data or theoretical analysis. Includes compilations of

* This is an edited, reformatted and augmented version of National Aeronautics and Space Administration Publication No. NASA/ TM□20240008449, dated Sepember 2024.

In: Applications of Optical Measurement Techniques …
Editor: Octavian Melina
ISBN: 979-8-90134-275-6

significant scientific and technical data and information deemed to be of continuing reference value.
NASA counterpart of peer-reviewed formal professional papers but has less stringent limitations on manuscript length and extent of graphic presentations.

TECHNICAL MEMORANDUM.
Scientific and technical findings that are preliminary or of specialized interest, e.g., quick release reports, working papers, and bibliographies that contain minimal annotation. Does not contain extensive analysis.

CONTRACTOR REPORT. Scientific and technical findings by NASA-sponsored contractors and grantees.

CONFERENCE PUBLICATION.
Collected papers from scientific and technical conferences, symposia, seminars, or other meetings sponsored or co-sponsored by NASA.

SPECIAL PUBLICATION. Scientific, technical, or historical information from NASA programs, projects, and missions, often concerned with subjects having substantial public interest.

TECHNICAL TRANSLATION.
English-language translations of foreign scientific and technical material pertinent to NASA's mission.

Specialized services also include organizing and publishing research results, distributing specialized research announcements and feeds, providing information desk and personal search support, and enabling data exchange services.

For more information about the NASA STI program, see the following:

Access the NASA STI program home page at http://www.sti.nasa.gov

Abstract

In support of the National Aeronautics and Space Administration Advanced Air Transonic Transport Truss-Braced Wing structural research, the Armstrong Flight Research Center Flight Loads Laboratory is performing pathfinder tests on a small, scaled test article (dubbed the mock Truss-Braced Wing) to study the application of a loads calibration on a truss-braced wing structure. The addition of a truss to the wing structure, as opposed to a purely cantilever beam setup, changes the strain-gage-based loads calibration techniques. In a truss-braced configuration, the internal shear and moments, inboard of the truss-wing connection, are not directly representative of the external aerodynamic loading as is the case for a traditional cantilever configuration. A 3-axis load cell was used at the wing-to-strut interface to estimate the load sharing between the two structural members. Crosstalk within the 3-axis load cell is added to the measurement uncertainties of the calibration effort, and the results, including crosstalk uncertainties, are presented. The calibration tests presented in this report are used as a pathfinder for developing an understanding how to perform loads calibration on a truss-braced wing structural configuration. The use of distributed fiber-optic strain sensors is leveraged to provide a near-full-span strain profile along the leading and trailing edge of the wing structure for strain-based loads calibration.

Nomenclature

AATT	Advanced Air Transonic Transport
AFRC	Armstrong Flight Research Center
CG	center of gravity
DTF	Displacement Transfer Function
FBG	fiber Bragg grating
FOSS	Fiber Optic Sensing System
OLEA	Operational Loads Estimation Algorithm
STA	wing structure station
tan	tangent
TRS	truss structure station
TBW	Truss-Braced Wing
NASA	National Aeronautics and Space Administration
d	distance, in
E	elastic modulus, $lbf - in^{-2}$
F	Force, lbf
h	distance between parallel optical fiber, *in*
I	second moment of area, in^4
lc	load cell

L	distance between sensing stations, *in*
M	bending moment, *in – lbf*
q	line load, *lbf – in–1*
V	shear load, *lbf*
x	forward and aft in local aircraft coordinate frame, *in*
y	right and left in local aircraft coordinate frame, *in*
z	up and down in local aircraft coordinate frame, *in*

Greek Symbols

Δ	difference
ε	normal strain, in/in
θ	wing twist angle, aircraft pitch angle, deg
σ	normal stress, $lbf - in^{-2}$
ϕ	wing slope angle, aircraft roll angle, deg
με	microstrain
Σ	summation

Subscripts

f, a	reference to forward location and aft location
i	reference station indices
LC	reference to load cell reference station indices
Mx, Mz	reference derived from moment about x or z
n	reference number stations
u, l	reference to upper surface and lower surface
x, y, z	reference orthogonal orientations
xm, ym, zm	reference of the measured orthogonal orientations

Introduction

The National Aeronautics and Space Administration (NASA) Advanced Air Transonic Transport (AATT) Program has developed a proof-of-concept scaled model for the Truss-Braced Wing (TBW). The TBW consists of high-aspect-ratio swept wings with an incident truss supported by the fuselage. The performance of such a TBW aircraft is being investigated by NASA by studying the potential aerodynamic efficiency and structural concept of the aircraft.

Multiple wind-tunnel tests have been performed by NASA (ref. 1), as shown in figure 1. Using this truss-braced configuration with a high-aspect-ratio swept wing, we believe that significant fuel savings could be achieved while simultaneously reducing emissions.

Figure 1. Truss-Braced Wing wind-tunnel test at NASA Langley Research Center (Hampton, Virginia).

It is imperative that test facilities engaged in structural testing enhance their abilities to conduct both instrumentation and loads testing of future wing concepts. The National Aeronautics and Space Administration (NASA) Armstrong Flight Research Center (AFRC) (Edwards, California) has an extensive history of performing loads calibrations on aircraft structures. The standard method for calibrating strain gages on a structure makes use of the Skopinski and Aiken method (ref. 2). The Skopinski and Aiken approach to loads calibration requires a large collection of applied loads and strain-gage data. The typical loads measured on a wing include shear, bending moment, and torque loads about a measurement station on the wing. The procedure requires the collection of applied loads and strain-gage data. The intent is to collect data from multiple load cases that vary the center of pressure on the wing surface.

Aiding to the development of new loads calibration techniques is the application of distributed strain sensing utilizing fiber Bragg gratings. A Fiber

Optic Sensing System (FOSS) can be used to interrogate optical fiber with a series (up to 1,000) of distributed fiber Bragg gratings (FBGs) to provide real-time strain data. During structural testing of the wing structure, the FOSS was able to simultaneously, and continuously, interrogate as many as eight 40-ft optical fibers with sensor locations spaced at 0.5-in intervals, providing nearly 8,000 strain sensor readings at a rate of 20 Hz (for more information regarding the FOSS technology and capabilities, see (ref. 3)). During testing of the mock TBW, there were a total of five fibers and approximately 1,500 unique strain sensors.

The FOSS provides 100 times the number of strain measurements at 1/100th the total sensor weight when compared to conventional electrical-resistance strain-gage implementations. The FOSS can operate in stand-alone mode (which is used for flight applications) or in remote-control mode (in which a laptop computer is connected to provide monitoring and control). A centralized software interface combines all functions into a suite of applications to fully utilize the FOSS.

Mock Truss-Braced Wing Description

The test article consists of a high-aspect-ratio rectangular planform with a 72-in half-span. The wing structure planform utilizes Z-shaped beams for forward and aft spars as well as rib stiffeners, and sheet metal for the upper and lower skins. The Z-shaped beams and the upper and lower skins are made of 1/16-in-thick 6061 aluminum. A cross-section view of the mock TBW is shown in figure 2. The strut structure consists of a ladder truss with square and rectangular tubes that is swept laterally by 10 deg, as shown in figure 3. The square and rectangular tubes are made of 1/16-in-thick 6061 aluminum. In addition, the truss is incident vertically by 11.7 deg into a swivel eyebolt at 35.8 in from the root, as shown in figure 4. The mock AATT Truss-Braced Wing test article is mounted on a strongback to ensure the degree of freedom (DOF) at the planform root is as rigid as possible for the test. The framing plate is secured into the strongback with two steel L-beams. The test article was mounted upside down to facilitate hanging loads on the test article to simulate lift, as shown in figure 5.

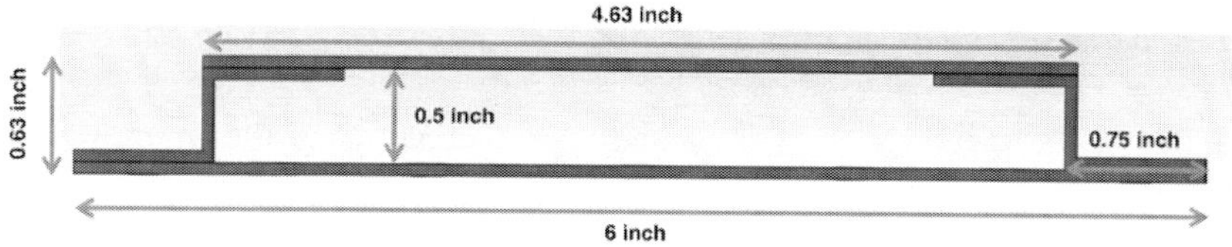

Figure 2. Mock Truss-Braced Wing cross-section view.

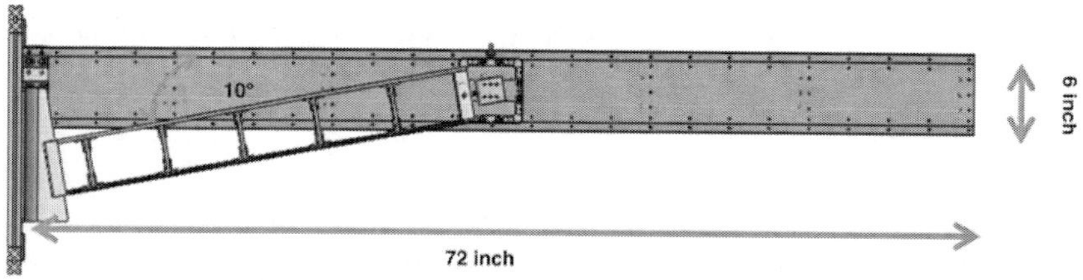

Figure 3. Mock Truss-Braced Wing (top view).

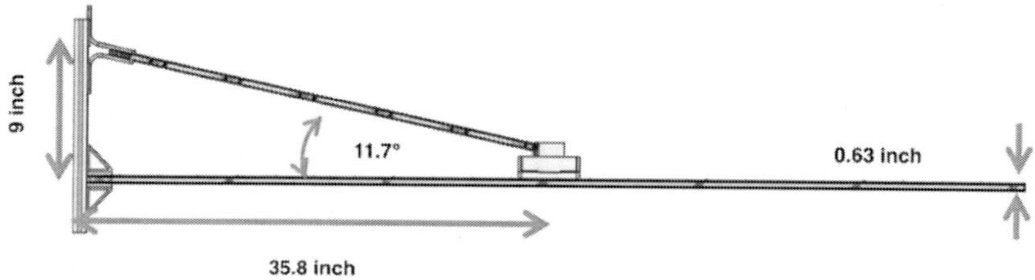

Figure 4. Mock Truss-Braced Wing (side view).

Figure 5. Mock Truss-Braced Wing on strongback.

Load Cases

A total of six load cases were applied during testing of the mock TBW structure, a root-biased load, mid-span-biased load, tip-biased load, uniformly distributed load, elliptical load and a 100-lb wingtip concentrated load. The total applied load over the full test series was limited to 200 lb of applied load for any given load case. Each load case utilized a combination of metallic slotted weights and shot bags. The wing structure was mounted upside down so that gravity would point in the lift direction. A sling system was used to hang the slotted weights underneath the wing structure, with clamps gripping the leading and trailing edge. Shot bags were placed on top of the ladder truss structure to simulate lift from the strut. The load case matrix containing the six applied load cases is shown in figure 6, and the loading locations are shown in figure 7. The application of the elliptical loading condition applied to the finite element model of the mock TBW structure is shown in figure 8, and the remaining figures of the finite element model load cases can be found in Appendix A (figures A1 to A5). An image of the mock TBW structure under test is shown in figure 9. The maximum load of each load case was held for approximately 30 s before the loads were removed. All data provided in this report are from the average sensor response during the 30-s hold.

Load case	Amount of weight added											Weight on structure (lbf)
	STA1	TRS2	STA2	TRS3	STA3	TRS4	STA4	STA5	STA6	STA7	STA8	
Root-biased load	40	10	40	10	40	10	50					200
Mid-biased load					20	5	25	25	25			100
Tip-biased load								25	25	25	25	100
Elliptical load	26	4	26	4	24	4	28	26	24	20	14	200
Uniform load	20	5	20	5	20	5	25	25	25	25	25	200
Tip load											100	100

Figure 6. Load case matrix.

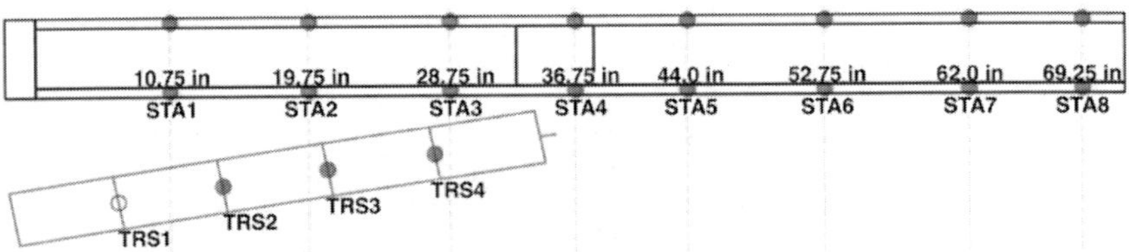

Figure 7. Loading stations.

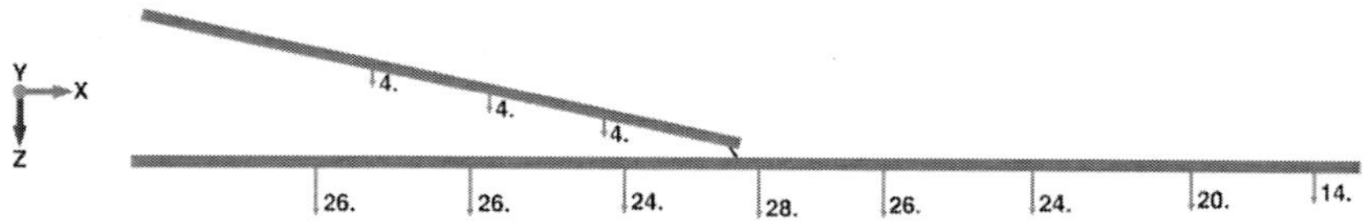

Figure 8. Finite element model of elliptical loading condition.

Figure 9. Elliptical loading condition.

Instrumentation

The mock TBW was instrumented with various types of instrumentation to characterize the structural response to applied load. Instrumentation included measurements of strain, displacement, angles, and internal structural load sharing at the wing-to-strut interface. A 3-axis load cell was installed between the main wing and strut interface, with an eyebolt utilized on the strut interface to release all rotational DOF, as shown in figure 10. Wing displacement measurements were made utilizing string potentiometers at three span stations along the leading edge, outboard the strut. Wing-twist-angle and slope-angle measurements were made using dual-axis inclinometers. Distributed strain measurements were obtained by distributed fiber Bragg gratings (FBGs) and an AFRC in-house-developed Optical Frequency Domain Reflectometry (OFDR) interrogator, referred to as the Fiber Optic Sensing System (FOSS). A typical AFRC-developed FOSS unit can simultaneously and continuously interrogate eight 40-ft optical fibers at 0.5-in intervals, providing nearly 8,000 strain sensor readings per scan at a typical rate of 20 Hz.

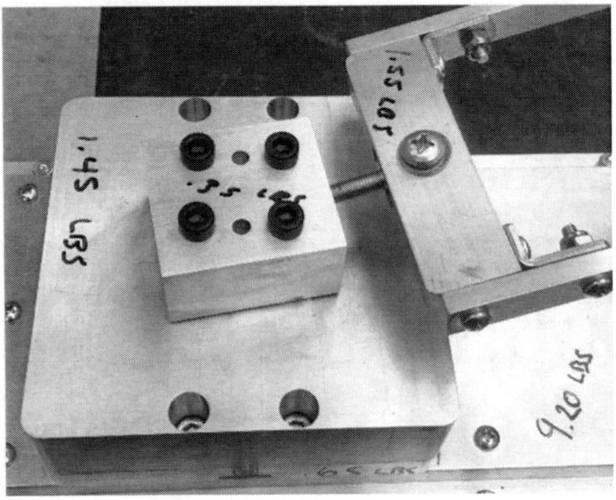

Figure 10. 3-axis load cell (dummy block) connecting wing structure to strut.

The FOSS was utilized as the only strain-sensing instrumentation. A 0.5-in-spaced distributed FBG on each fiber-optic channel was routed along the leading and trialing edge of the wing structure along with ±45 deg patterns along the upper and lower surface. The FBGs routed in the ±45-deg directions are primarily sensitive to torsion. The FBGs routed along the leading and trailing edges are sensitive simultaneously to out-of-plane bending as well as in-plane bending. Due to the mounting of the load cell to the wing structure, the fiber-optic pattern on the lower surface was not continuous along the span but, instead, was split into an inboard and an outboard region. Distributed FBGs were also installed on the top and bottom of the strut structure. Figure 11 shows the FOSS layout pattern on the upper and lower surfaces.

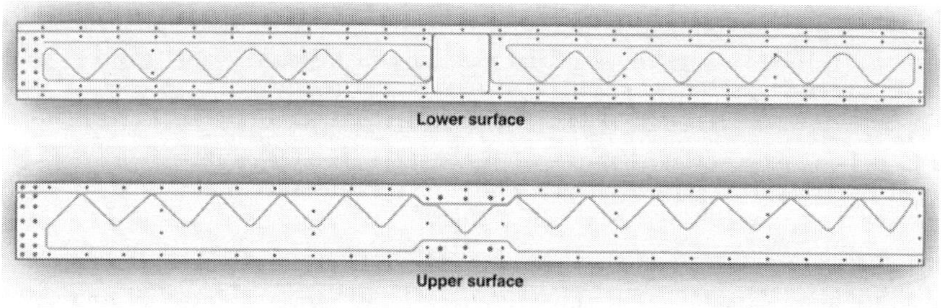

Figure 11. Fiber Optic Sensing System path on the mock Truss-Braced Wing (lower and upper surfaces).

The FOSS was also applied on the strut structure along the center of the forward and aft beam and on the upper and lower surface. Figure 12 shows the FOSS layout pattern on the strut structure.

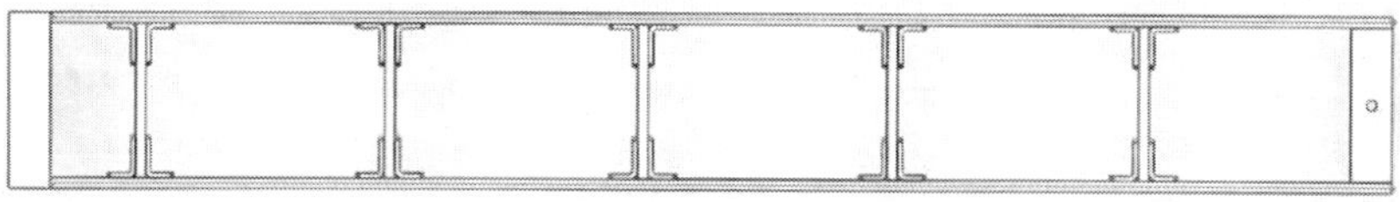

Figure 12. Fiber Optic Sensing System path on the mock Truss-Braced Wing, strut structure.

Operational Loads Estimation Algorithm

The ability to monitor hundreds or thousands of strains has led to the development of shape and load algorithms that leverage new technology. Previous work at AFRC by Dr. William L. Ko, et al. (refs. 4-6) has shown the benefits of using a displacement transfer function to calculate the deformed shape of a wing under external loading. A large-scale load test of the Global Observer wing (AeroVironment, Monrovia, California) validated the shape algorithms (ref. 7). Richards and Ko (refs. 8 and 9) have patented a method for calculating operational loads, but sensor-to-sensor measurement variations can make the method challenging to implement. Validation tests confirming the bending moment calibrations were demonstrated by utilizing distributed fiber-optic sensors and the method proposed by Richards and Ko (ref. 10). The Richards and Ko method relies on taking a numerical derivative of bending moments with respect to span to estimate shear load. Though the equations are mathematically correct, spatial measurement variations make the output less accurate. It is generally advisable to avoid numeric differentiation whenever possible (ref. 11) because small inaccuracies in measurements can produce large errors as the length domain converges to a small distance. An approach by Pak (ref. 12) uses the deformed shape from modal testing to calculate the accelerations and velocities. The modal deformed shapes are then used to estimate the loads on the wing.

This report describes an algorithm developed at AFRC that allows the calculation of shear and bending moment distribution along the wingspan, utilizing distributed strain measurements to compute wing bending shape and loading information. A brief formulation of the algorithm is described herein.

The typical local coordinate frame for an aircraft defines the local X-axis to be aligned forward and aft, and the rotation about the X-axis (referred to as the roll angle) is defined as ϕ. The change in slope of the aircraft wing structure due to structural deformation will also be defined as ϕ. Orthogonal to the local aircraft X-axis and on the horizontal plane of reference is the local aircraft Y-axis that is aligned left and right, and the rotation about the Y-axis is referred to as the pitch angle and is defined as θ. In this report, the change in pitch of the aircraft wing structure will also be referred to as twist angle and will also be defined as θ. The local aircraft Z-axis is aligned vertically in the up and down direction. An illustration of a typical aircraft local coordinate system frame is shown in figure 13.

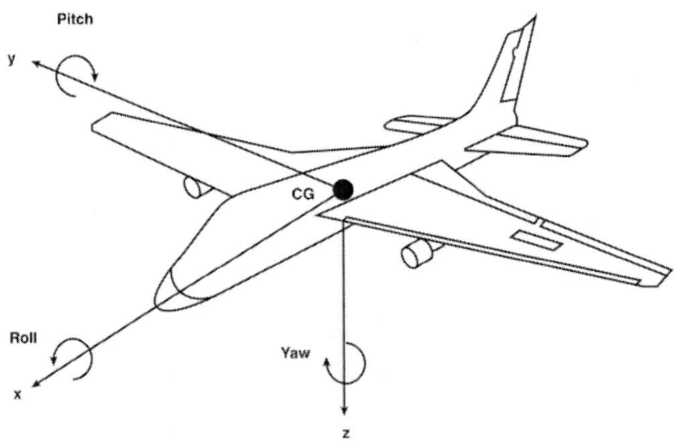

Figure 13. Aircraft local coordinate system frame.

The process of calibrating strain gages to map to bending moments requires determining the cross-sectional properties of this structure and is accomplished via structural testing. A calibration load case is applied to the test article with a known applied load, and the strain response is monitored. All strain measurements throughout this test series are generated from the FOSS. The strain-based load prediction algorithm, also known as the Operational Loads Estimation Algorithm (OLEA), is derived from the Classical Beam Theory (Euler-Bernoulli) for several beam configurations including cantilever beams, which are analogous to the wing spar structure of the aircraft. The Classical Beam Theory can be used to relate the loading of a structure to the theoretical curvature, slope, and deflection of a cantilever beam. The general differential equation that describes the relationship between a line loading function and the resulting shape of a beam is shown as eq. (1) (ref. 11):

$$\frac{d^4 z}{dy^4} = \frac{q}{EI_x} \tag{1}$$

where z is the vertical displacement of a beam as a function of distance from the root; q is the line loading function (load per unit length) analogous to the lift generated by the wing; E is the elastic modulus; and Ix (or Iz) is the second moment of area that is calculated from the centroidal axis perpendicular to the applied loading.

The general differential equation that describes the relationship between vertical shear and the resulting shape of the beam is shown as eq. (2) (ref. 11):

$$\frac{d^3 z}{dy^3} = \frac{V}{EI_x} \tag{2}$$

where the V is the vertical shear along the span of the structure. For cantilevered structures, the vertical shear is the summation of vertical applied load starting with zero load at the wingtip and resulting with maximum total applied load at the root.

For the purposes of implementing the strain-based OLEA, the equation of interest (shown as eq. (3)) is the moment-curvature relationship of the classical beam differential equation that states that the curvature, which is the second derivative of displacement (z) of a beam, is proportional to the applied bending moment (ref. 11).

$$\frac{d^2 z}{dy^2} = \frac{M_x}{EI_x} \quad (3)$$

Since the bending moment cannot be measured directly, strain sensors are placed on the upper and lower surfaces to quantify the bending moments. For out-of-plane bending or bending strain due to bending moment about the global X-axis, the bending strain (εMx) is defined as the upper surface strain (εu) subtracted by the lower surface strains (εl), as shown in eq. (4). For in-plane bending or bending strain due to bending moment about the global Z-axis, the bending strain (εMz) is defined as the forward surface strain (εf) subtracted by the aft surface strains (εa), as shown in eq. (5). The rest of this section will focus on out-of-plane bending and shear; however, applications to in-plane bending can used. An illustration of out-of-plane bending is shown in figure 14, and an illustration of in-plane bending is shown in figure 15.

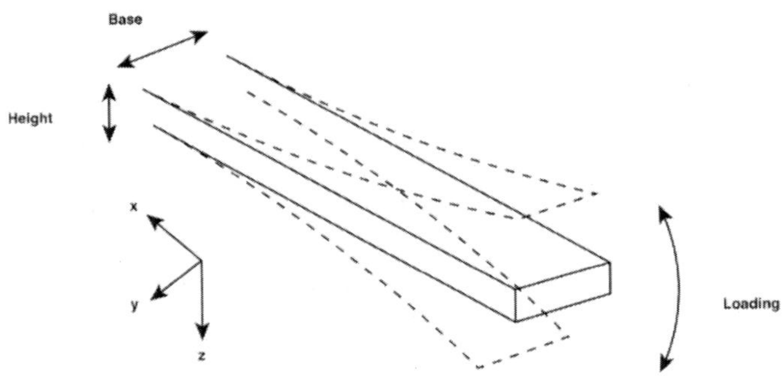

Figure 14. Out-of-plane bending.

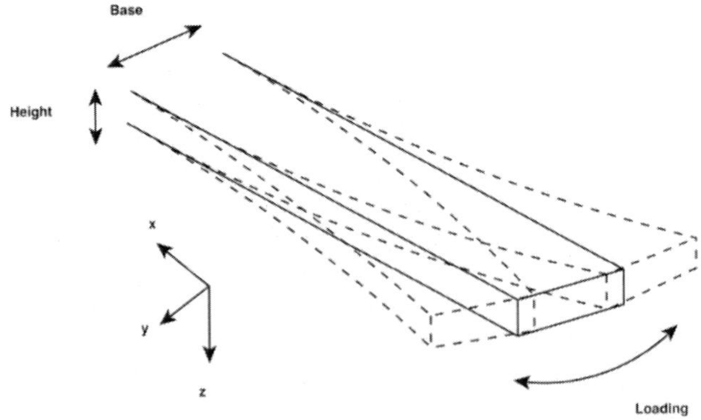

Figure 15. In-plane bending.

The technique of relating bending moments to the difference in axial strain between different surfaces is similar to the process of utilizing a four-active-arm Wheatstone Bridge Circuit (full- bridge circuit) set-up in a moment bridge configuration, where two foil gages are placed on one surface, and the other two foil gages are placed on a separate surface (ref. 2). The bending strain equations about the X- and Z-axis is provided in eqs. (4) and (5).

$$\varepsilon Mx = \varepsilon u - \varepsilon l \tag{4}$$

$$\varepsilon_{Mz} = \varepsilon_f - \varepsilon_a \tag{5}$$

The differential bending stresses (σx and σz) are defined as the bending moment multiplied by the distance between paired strain sensors (h1 or h2) relative to the bending plane of interest, divided by the second moment of area (Ix or Iz) about the axis of interest (ref. 11). The bending stress is also related to bending strain by multiplying the elastic modulus (E), as shown in eqs. (6) and (7).

$$\sigma_{M_x} = \frac{M_x h_1}{I_x} = E\left(\varepsilon_u - \varepsilon_l\right) \tag{6}$$

$$\sigma_{M_z} = \frac{M_z h_2}{I_z} = E\left(\varepsilon_f - \varepsilon_a\right) \tag{7}$$

By rearranging the terms in eqs. (3), (4) and (5) we see that the bending moment can be related to bending strains, as shown in eqs. (8) and (9), where the elastic modulus (E); second moment of area (I_x and I_z); and distance between strain sensors (h_1 and h_2) are all constants. The individual constants do not need to be solved independently but, rather, are calculated as a combined value carried forward to map bending strains to bending moments. The process of mapping bending strain to bending moments is referred to as the load transfer function (refs. 8-10), shown in figure 16.

$$M_x = \varepsilon_{Mx}\left(\frac{EI_x}{h_1}\right) \tag{8}$$

$$M_z = \varepsilon_{Mz}\left(\frac{EI_z}{h_2}\right) \tag{9}$$

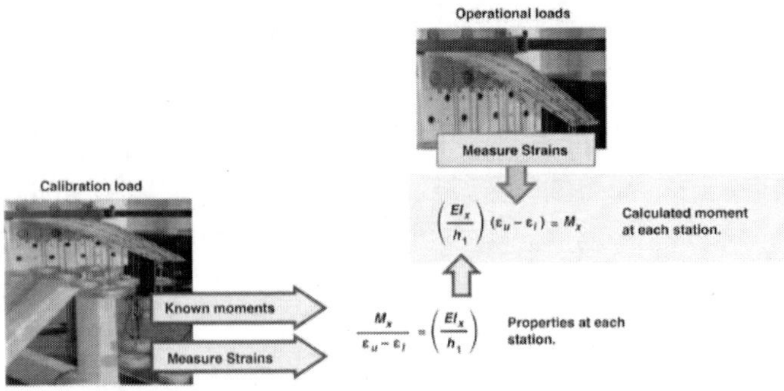

Figure 16. Load transfer function implementation process (ref. 10).

Since the FOSS strain measurements are comprised of discrete sensors, it is possible to numerically integrate the curvature in eq. (3) utilizing the trapezoidal rule. The distributed slope (ϕ) of the structure can be estimated by eq. (10):

$$\frac{dz}{dy} = \tan(\phi) \approx \phi_n = \phi_0 + \sum_{i=1}^{n-1} \left(\left(\frac{M}{EI_x} \right)_i + \left(\frac{M}{EI_x} \right)_{i+1} \right) \frac{\Delta L_i}{2} \tag{10}$$

where $\tan(\phi) \approx \phi$ for small angles, ϕ_i is the slope angle in the roll orientation at the ith sensing element, with the nth station representing the last sensing station at the tip, ϕ_0 is the root slope, and Δ_L is the distance between the sensor stations. The small angle approximation for the tangent function produces a relative error of less than 1 percent for angles less than 10 deg (0.175 rad), and less than a 1-percent error for the sine function with less than 14 deg (0.244 rad). The small angle approximation also linearizes the relationship between loads and displacement. The vertical displacement (Z_n) at each sensing station can be approximated by integrating the slope equation. The process of determining the vertical displacement at any sensing station is called the Displacement Transfer Function (DTF) and can be determined by eq. (11).

$$z_n = z_0 + \sum_{i=1}^{n} \sin(\phi_i) \Delta L_i \tag{11}$$

Using small angle approximation $\sin(\phi) \approx \phi$, the calculation for vertical displacement becomes eq. (12).

$$z_n = z_0 + \sum_{i=1}^{n} \phi_i \Delta L_i \tag{12}$$

An illustration of the OLEA flowchart is shown in figure 17. The OLEA utilizes the estimated shape (Zn) from eq. (12); the calibrated section properties (EIx) from eq. (8); and an initial guess of the shear loads as input. For the first iteration of OLEA, a small-magnitude uniform shear load is recommended. The guess shear load is then assigned to a temporary shear load variable for further processing. The shear load variable divided by the calibrated section properties are then integrated twice with respect to span to obtain a variable mathematical model of the deflection. The model deflection and the DTF deflections are then compared with one another to determine the error between the two distributed shapes. If the deflection error is not within a specified tolerance, the shear load estimate is updated proportionally to the amount of spanwise deflection error. The updated shear load variable is checked to ensure that the results are consistent with a cantilever configuration. The cantilever check ensures the shear load at the tip is set to zero and that each shear load sensing station ordered from root to tip is greater than or equal to the next sensing station. The resulting shear load estimate is then sent in a loop for further reprocessing to obtain an updated model deflection. This processing loop continues until the deflection error is within a specified tolerance or until a specified number of iterations are completed. Once the deflection error is within tolerance, the shear load estimate is further processed to obtain an estimate for bending moment. The output of the OLEA is the estimated spanwise shear load and bending moment. For sequentially timed calculations of load, we recommended that the output results from the previous time frame (time_{j-1}) be fed to the current initial estimate for shear load to reduce the computation time, enabling near-real-time calculation. The results presented in this report, however, represent the average response of each sensor during maximum loading, so no time optimization was applied.

Due to the interaction between the truss to the main wing section, the OLEA process is split into two sections: one section outboard of the load cell and another section inboard of the load cell. The outboard process was unmodified. The inboard section was processed as a cantilevered structure and assumed zero shear up to the load cell. After the OLEA processing of the inboard section was completed, an additional offset was included to account for the shear calculated from the outboard section and to account for the load cell output.

Loads Calibration

The focus of this report is on the calibration of the wing structure; therefore, all the calibrations shown in this report have focused on the wing structure and not the strut. A similar analysis, however, may be performed on the strut using the same techniques described in this report.

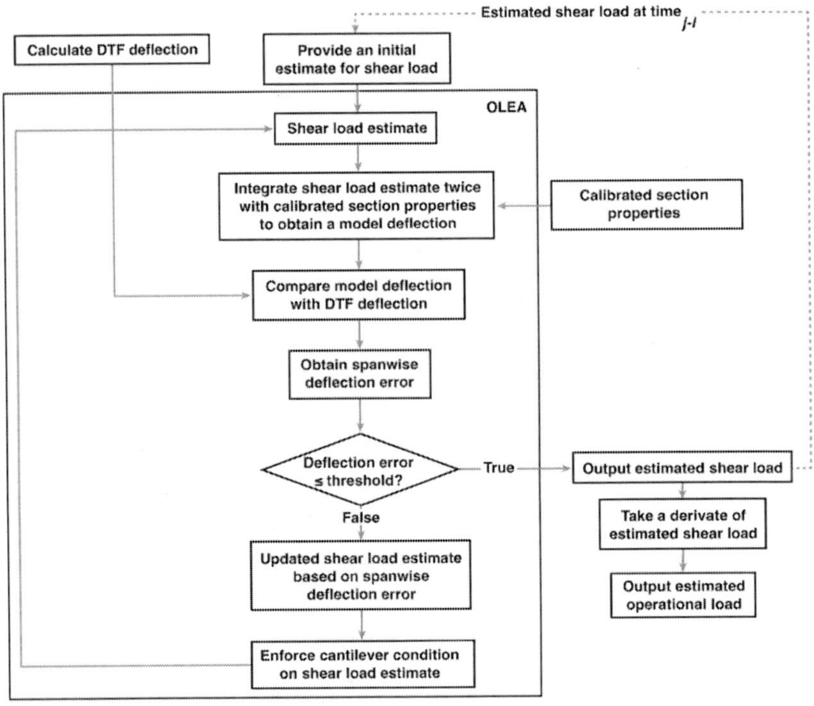

Figure 17. Operational loads estimation algorithm flowchart.

Out-of-Plane Calibration

To develop a free-body diagram for the mock TBW, measurements from the load cell that connect the wing structure to the strut must be taken into account. The load cell was installed off-center of the neutral axis (dy and dz) of the wing structure. As a result, bending moments that are caused by the load cell offset (MLC) need to be accounted for in addition to the load cell force-measurement output. The equation for the bending moments caused by the load cell offset is provided in eq. 13. Figure 18 shows the free-body diagram and the bending moment caused by the load cell offset, where Fz and Fy are the load cell outputs.

$$MLC = dzFy - dyFz \tag{13}$$

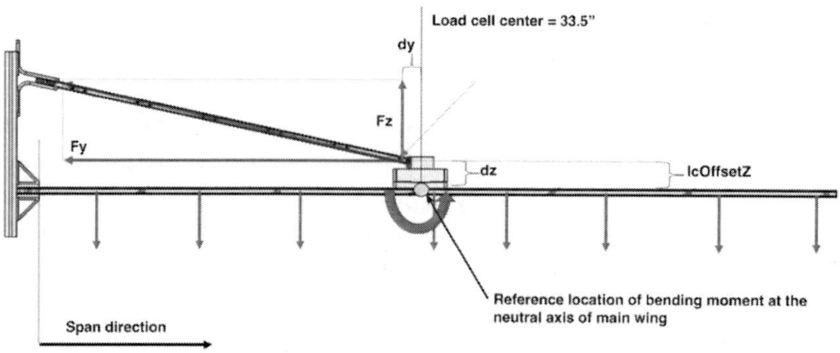

Figure 18. Mock Truss-Braced Wing free-body diagram for out-of-plane bending, with the structure mounted upside down to facilitate hanging weights to simulate lift loads.

An important property of the load cell that needs to be monitored is the crosstalk along the perpendicular sensing axis. The load cell used in the testing of the mock TBW test series had a stated 2-percent crosstalk effect on Fx and Fz due to loads generated in Fy; therefore, Fz is based on the measurements (Fym and Fzm) plus a 2-percent uncertainty of the perpendicular measurement. The calculated forces in the X- and Z-axis is presented in eqs. (14) and (15). For out-of-plane bending, the measurements in the forward and aft direction (Fx) are omitted.

$$F_z = F_{zm} \pm 0.02(F_{ym}) \tag{14}$$

$$F_y = F_{ym} \pm 0.02(F_{zm}) \tag{15}$$

Due to the shallow angle of the strut relative to the wing structure, F_y is on the order of five times larger than F_z. The crosstalk between F_y and F_z cause a measurement uncertainty of ±10 percent in F_z, while F_y had an uncertainty of 0.4 percent due to the crosstalk. The uncertainty in F_z is included in the out-of-plane bending moment diagrams since the uncertainty produces a wide range of possible bending moments inboard of the load cell. The outboard section of the wing, therefore, was used to determine the calibration parameters, and the calibration parameters were carried over to the inboard section since the wing cross section was uniform along the entire length. We recommend, whenever possible, to calibrate the wing and strut independently to avoid errors due to measurement uncertainties introduced when testing the structure as an assembly.

The out-of-plane bending moment calibration load case was based on the wingtip concentrated load case. In this load case, 100 lb was applied at the outboard loading station. The strain response from the wingtip concentrated load case is shown in figure 19 and shows that the strain along the top and bottom surfaces are plotted against the span. There were approximately 142 unique strain-measurement stations on both the upper and lower sensing fiber. The measured strain distribution along the bottom surface was discontinuous at the load cell interface due to the footprint of the load cell and how it was bolted flush to the wing surface; therefore, bending strain is not obtainable along the footprint of the load cell. Vertical-dash lines are shown in figure 19 to show the load cell location relative to span.

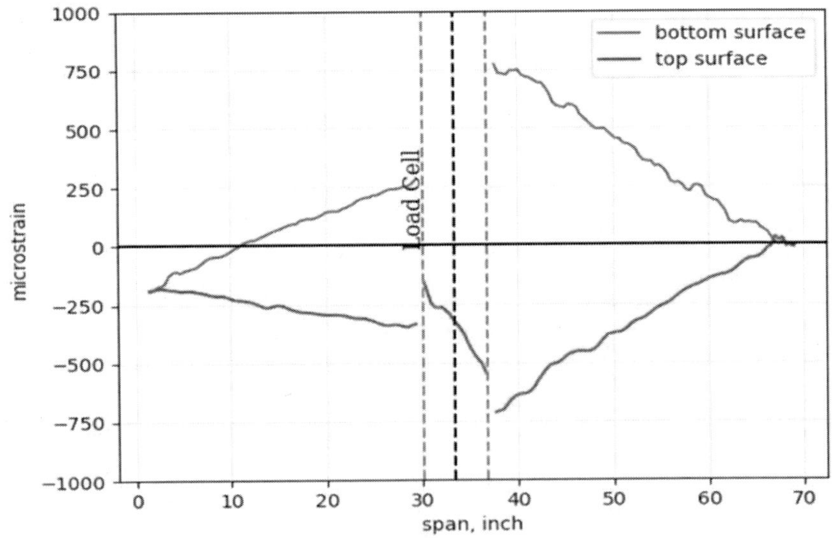

Figure 19. Fiber Optic Sensing System strain response to a concentrated wingtip load.

An example mapping between the applied bending moment and the unprocessed bending strain ε_{Mx} (see eq. (4)) response is shown in figure 20. Within figure 20, the left-vertical axis is the bending strain response (blue curve) in microstrain, and the right-vertical axis is the bending moment put into the test article in in-lbf (red curve). The results for the additional ±2-percent crosstalk are also shown in figure 20, with the orange-shaded region representing the range of values. The vertical-dash lines represent the load cell footprint where the bending strains were not obtained.

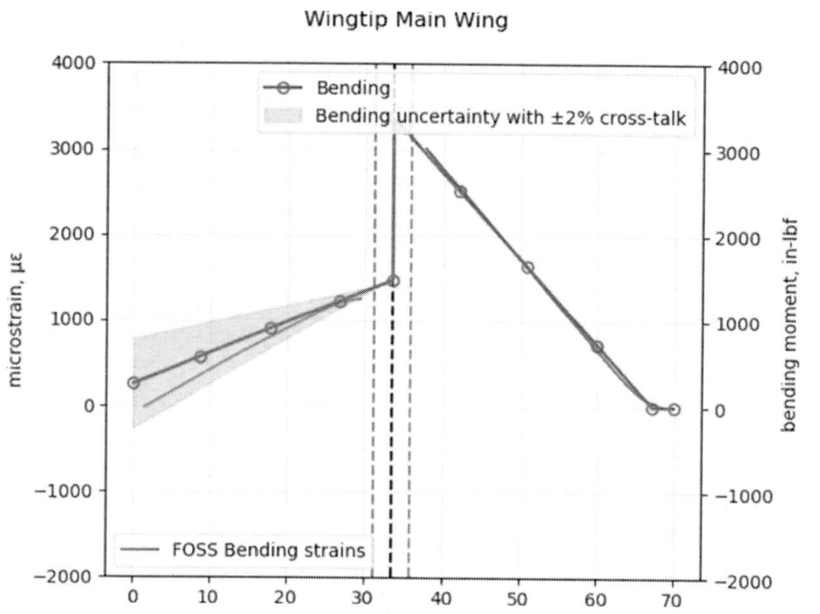

Figure 20. Out-of-plane bending moment diagram for concentrated wingtip load case.

Figure 20 shows that the bending strain response and the applied bending moment align well along the outboard portion of the wing beyond the load cell. The bending strain measurements inboard of the load cell are within the tolerance of the output of the 3-axis load cell. The strain measurements and bending moments outboard of the load cell are used for calibration of the operational loads estimation algorithm, since the inboard portion has some uncertainties due to crosstalk between the sensing axes. The average calibration value that maps the bending strain to bending moment is approximately EIx/h1 − 2.02 in-lbf per microstrain.

In-Plane Calibration

Due to the interaction between the wing and the swept strut structure, there exists an in- plane bending moment even though there were no external loads applied in the fore or aft direction. The in-plane bending moment is determined by the fore or aft output of the 3-axis load cell measurement (F_{xm}). As mentioned in the previous section, "In-plane Calibration," the crosstalk between forces in the fore or aft direction, and the inboard or outboard direction need to be accounted for. The measurement uncertainty utilized to determine F_x are provided in eq. (16).

$$F_x = F_{xm} \pm 0.02(F_{ym}) \tag{16}$$

The strain response from the FBGs on the forward and aft fibers routed on the upper surface is utilized in determining the in-plane bending strains, ε_{Mz}, in eq. (5) and is shown in figure 21.

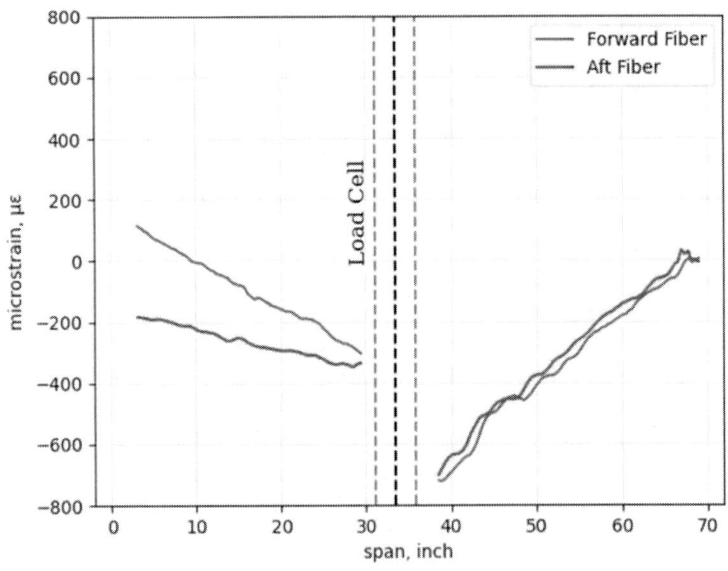

Figure 21. In-plane Fiber Optic Sensing System strain response for concentrated wingtip load case.

Figure 21 shows the vertical axis is the measured strain from the FOSS, and the horizontal axis is the span distance. There were approximately 142 unique strain measurement stations on both the forward and aft sensing fiber. As mentioned in the previous section, "In-plane Calibration," the measured strain distribution along the span was discontinuous at the load cell interface due to the footprint of the load cell and how it was bolted flush to the wing surface. Bending strain, therefore, is not obtainable along the footprint of the load cell. Vertical-dash lines are shown in figure 21 and represent the load cell location relative to span.

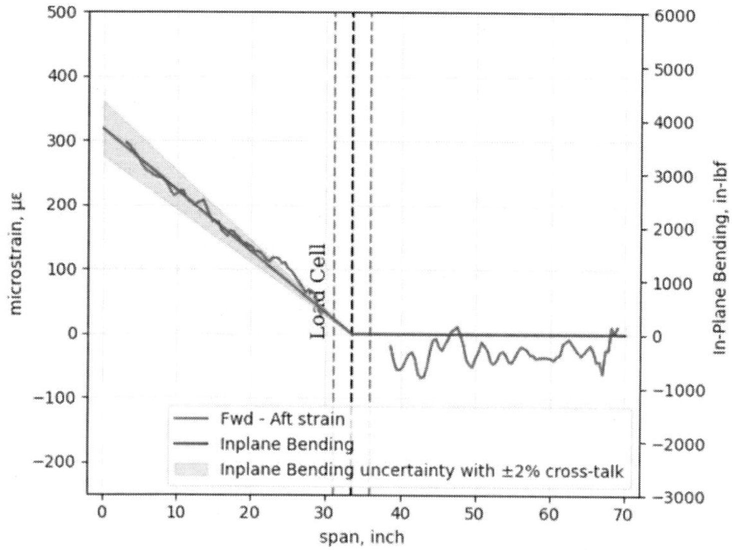

Figure 22. In-plane bending moment diagram for concentrated wingtip load case.

An example mapping between the applied bending moment and the unprocessed bending strain εMz, (see eq. (5)) response is shown in figure 22. Within figure 22, the left-vertical axis is the bending strain response (blue curve) in microstrain, and the right-vertical axis is the in-plane bending moment put into the test article in in-lbf (red curve). The results for the additional ±2-percent crosstalk are also shown in figure 22, with the orange-shaded region representing the range of values. The vertical-dash lines represent the load cell footprint where the bending strains were not obtained. The focus of figure 22 is on the bending moments inboard of the load cell since there is no external loading outboard of the load cell. The average calibration value that maps the bending strain to bending moment is approximately EIz/h2 = 11.6 in-lbf per microstrain. It is noted that the strain output in the in-plane direction has relatively low output when compared to the out-of-plane loading due to the much larger second moment of area of the structure in the in-plane direction, causing the structure to be much stiffer in this loading direction. The strain output appears much noisier in the in-plane direction.

Shear and Bending Loads Estimations

Out-of-Plane

The FOSS out-of-plane bending strains (eq. (4)) were then processed into a linearized deflection profile using eqs. (10) and (12). The linearized deflection along with section properties (EIx) were processed using OLEA (figure 17). The output of OLEA provides a qualitative estimate of the shear loads and bending moments. The bending moment component of the OLEA process for the elliptical load case is shown in figure 23and shows that the blue curve is the spanwise FOSS-derived OLEA bending moment estimation, the red curve is the load-cell-derived bending moment, and the ±2-percent crosstalk of the load cell is shown as the orange-shaded region representing the range of values. The OLEA out-of-plane bending results for the remaining five load cases are provided in Appendix B (figures B1 to B5).

The shear component of the OLEA process is shown in figure 24, where the blue curve is the spanwise FOSS-derived OLEA shear estimation, the red curve is the load-cell-derived shear, and the ±2-percent crosstalk of the load cell is shown as the orange-shaded region representing the range of values. Figure 24 shows that the qualitative profile of the shear load is captured by the OLEA. It is also apparent that the OLEA technique is not able to provide the piecewise linear curves that satisfy the true applied shear load curve calculated from the load cells, as shown in figure 24. This result is because vertical deflection is used as the optimization parameter for convergence in the OLEA feedback loop. The vertical deflection is a smooth, continuous curve, and by utilizing the spanwise error calculations of a smooth curve to update the shear loads, a smooth shear load curve is produced. The inability of the OLEA to predict sharp, piecewise linear curves, produces comparatively large errors at kinked locations for the load cell shear calculations. It is believed that the OLEA calculations would produce reduced errors for shear estimates in a flight environment in which the applied loads are distributed pressure loads versus discrete span locations with hanging weights. The OLEA out-of-plane shear results for the remaining five load cases are provided in Appendix C (figs, C1 to C5).

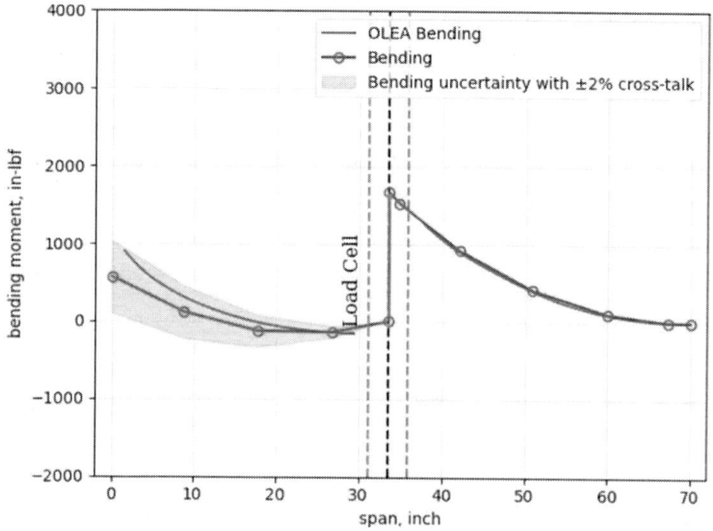

Figure 23. Out-of-plane bending moment diagram for elliptical load case.

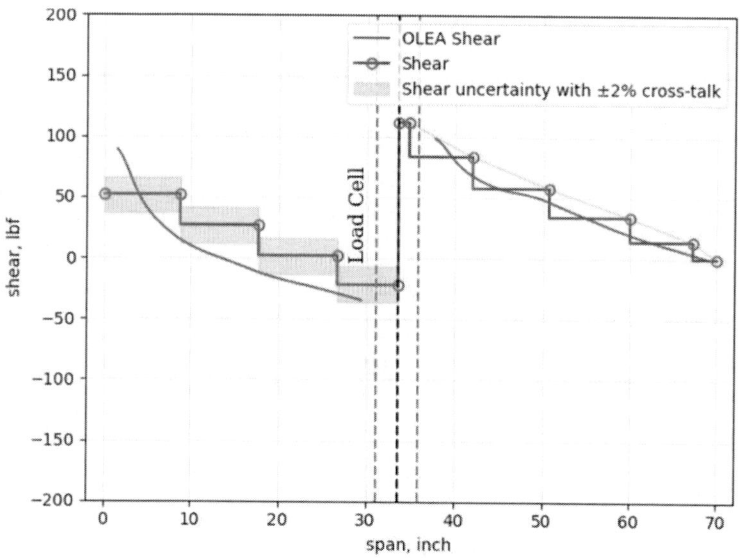

Figure 24. Operational Loads Estimation Algorithm out-of-plane shear diagram for elliptical load case.

In-Plane

The FOSS in-plane bending strains (eq. (5)) were then processed into a linearized deflection profile using eqs. (10) and (12). The linearized deflection along with section properties (EI_z) were processed using the OLEA (figure 17). The bending moment component of the OLEA process for the elliptical load case is shown in figure 25, where the blue curve is the spanwise FOSS-derived OLEA bending moment estimation, the red curve is the load-cell-derived bending moment, and the ±2-percent crosstalk of the load cell is shown as the orange-shaded region representing the range of values. The OLEA in-plane bending results for the remaining five load cases are provided in Appendix D (figures D1 to D5).

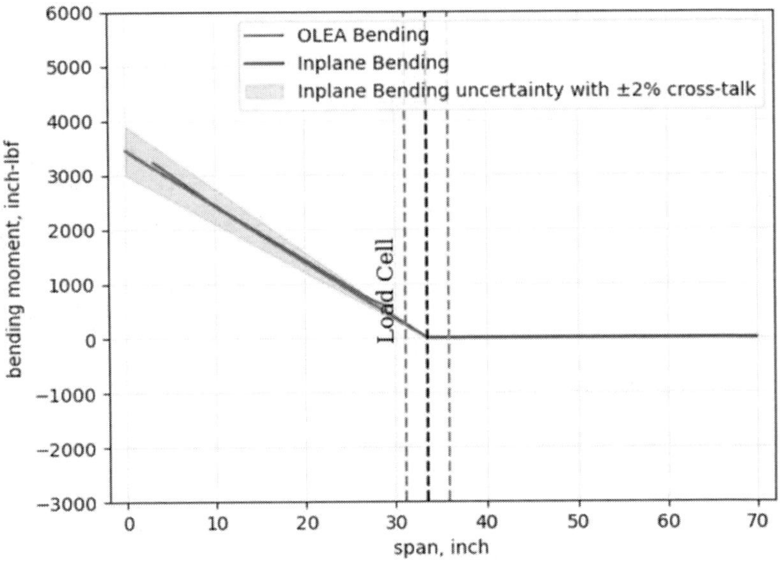

Figure 25. Operational Loads Estimation Algorithm in-plane moment diagram for elliptical load case.

The shear component of the OLEA process is shown in figure 26, where the blue curve is the spanwise FOSS-derived OLEA shear estimation, the red curve is the load-cell-derived shear, and the ±2-percent crosstalk of the load cell is shown as the orange-shaded region representing the range of values. Figure 26 shows that the qualitative profile of the shear load is captured by the

OLEA. The OLEA in-plane shear results for the remaining five load cases are provided in Appendix E (figures E1 to E5).

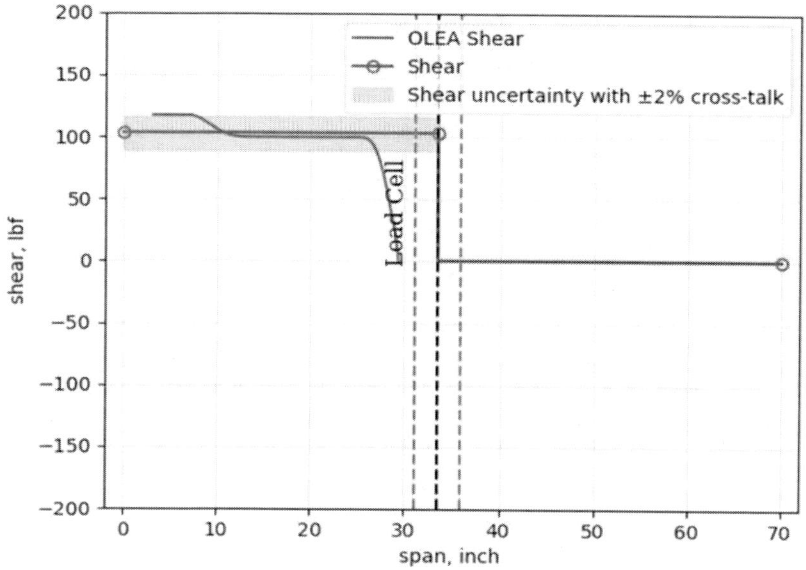

Figure 26. Operational Loads Estimation Algorithm in-plane shear diagram for elliptical load case.

Out-of-Plane Bending Shape Estimation

The FOSS strain data along the upper and lower surfaces were processed to estimate the out-of-plane displacement using eqs. (6), (10), and (11) for all load cases. The FOSS strain distribution utilized in the shape estimate is shown in figure 27. The shape estimate for the elliptical load distribution case is shown in figure 28. Within figure 28, the blue curve represents the FOSS displacement outputs, and the red circles represent co-located string potentiometers. As mentioned throughout the report, the strain information under the load cell region was discontinuous, causing some deviations in the shape output when compared to the string potentiometers. To compensate, a ramp function was used to bridge the strain distribution from the inboard section to outboard section under the load cell. In general, if accurate shape is desired, continuous distributed strain measurements from the root to the tip of the structure of interest is advised. The displacement estimates for the remaining load cases are provided in Appendix F (figures F1 to F5).

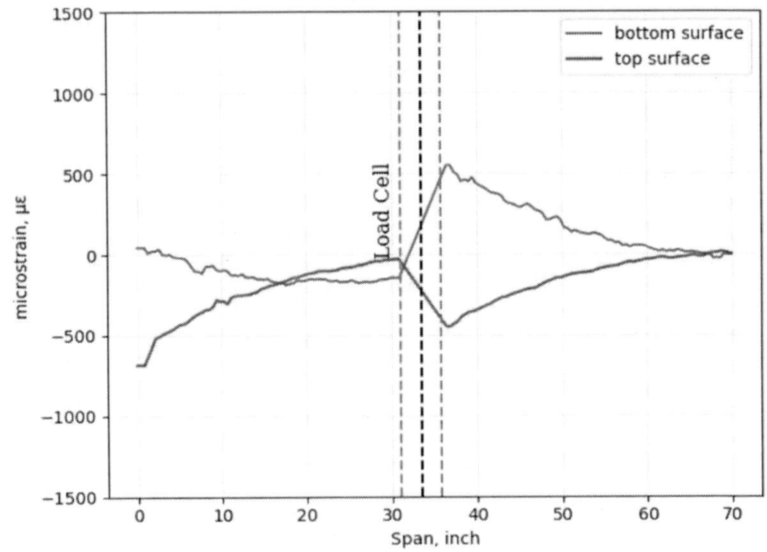

Figure 27. Fiber Optic Sensing System strain distribution for elliptical load case.

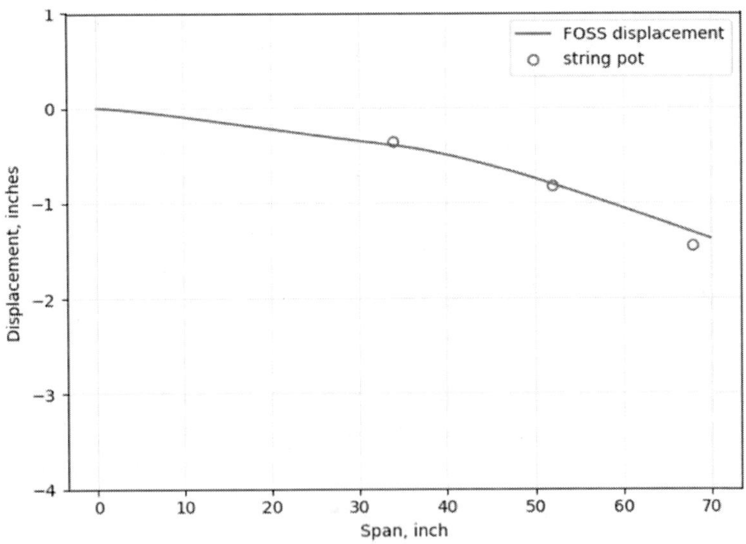

Figure 28. Out-of-plane displacement for elliptical load case.

Conclusion

The loads calibration research performed on the mock Truss-Braced Wing test article highlights the usefulness of utilizing distributed strain sensing technology, such as the National Aeronautics and Space Administration Armstrong Flight Research Center Fiber Optic Sensing System. Distributed surface strain measurements can be utilized with the Fiber Optic Sensing System to estimate the stiffness characteristics of a wing without using a finite element model. These distributed strain measurements can, in turn, provide a fully defined bending moment and an estimate of the shear load diagrams. Due to the measurement uncertainties of the selected 3-axis load cell used in these test series to monitor the load sharing between the wing and strut, the accuracy of the bending and shear diagrams cannot be reported for this test series.

Though the applied operational loading information provided in this report was shown to be more qualitative than quantitative, the distributed load sensing measurement can continue to improve with advancements in sensing technology, increased computer performance, and the potential incorporation of improved algorithms or machine-learning algorithms. The use of a 3-axis load cell with crosstalk makes the determination of the true internal loads difficult to obtain; therefore, estimates related to accuracy of the FOSS-derived loads is qualitative. The development of smart adaptive structures could lead to lighter-weight structures and, thus, improved structural safety and fuel efficiency.

Appendix A: Load Cases

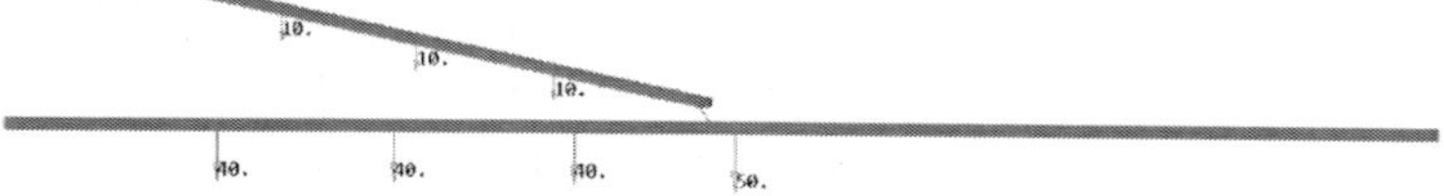

Figure A1. Finite element model of root-biased loading condition.

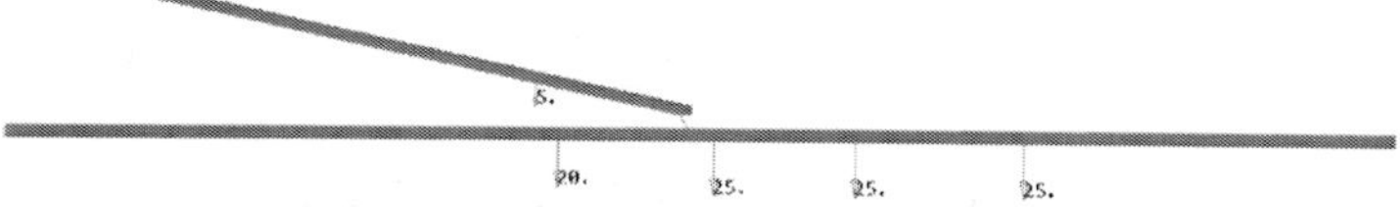

Figure A2. Finite element model of mid-span-biased loading condition.

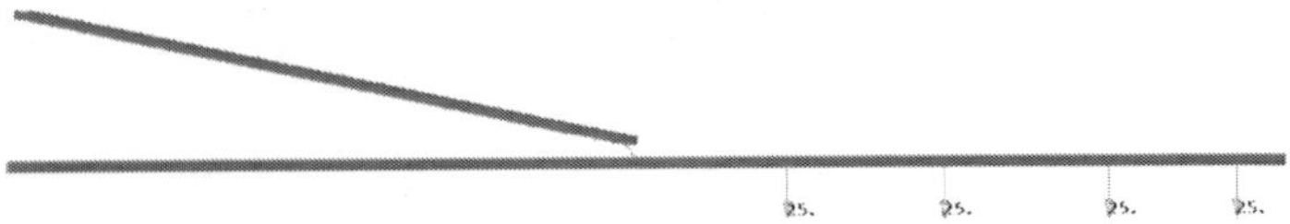

Figure A3. Finite element model of tip-biased loading condition.

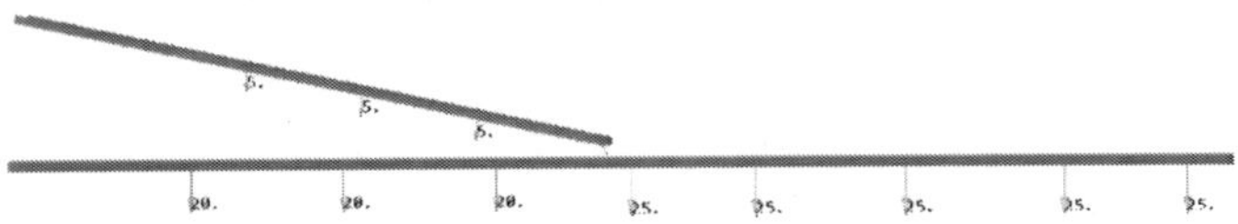

Figure A4. Finite element model of uniform loading condition.

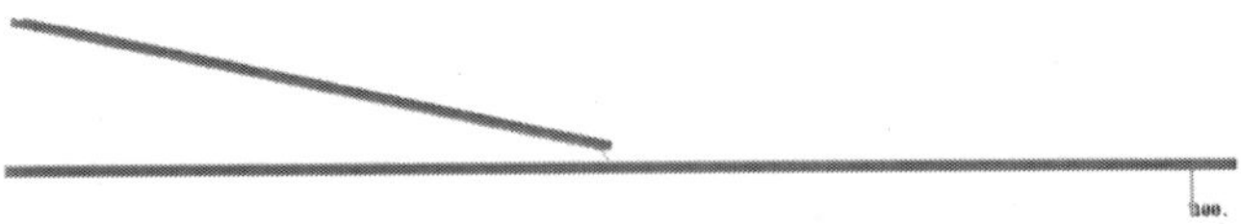

Figure A5. Finite element model of tip loaded condition.

Appendix B: Out-of-plane Bending Diagrams

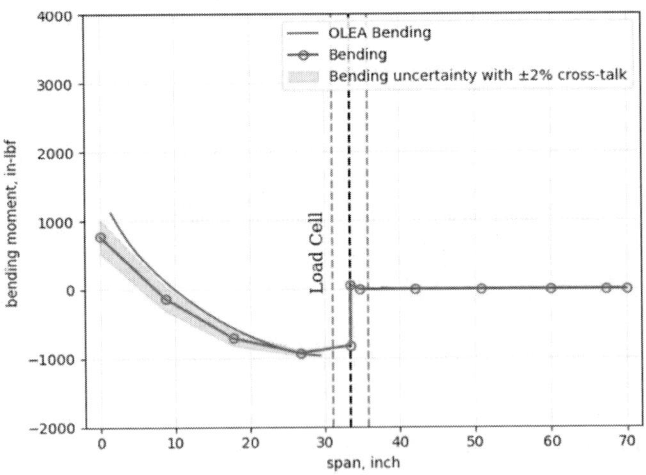

Figure B1. Out-of-plane bending moment diagram for root-biased load case.

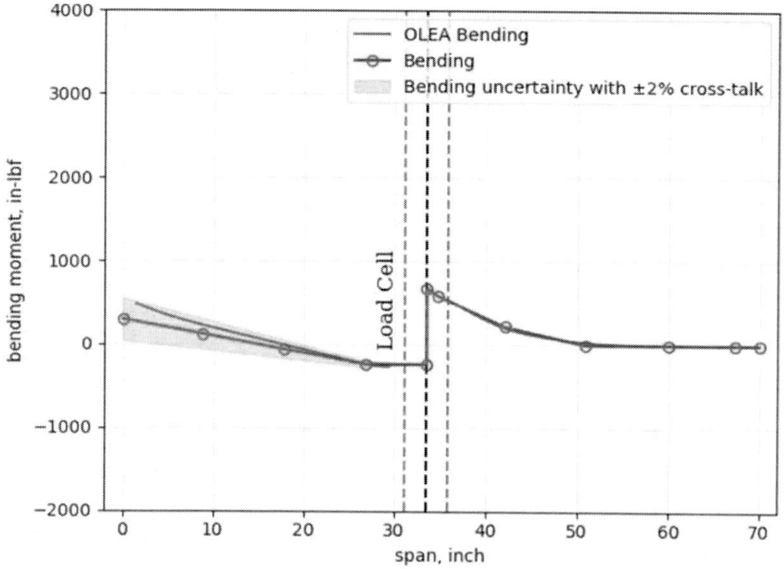

Figure B2. Out-of-plane bending moment diagram for mid-span-biased load case.

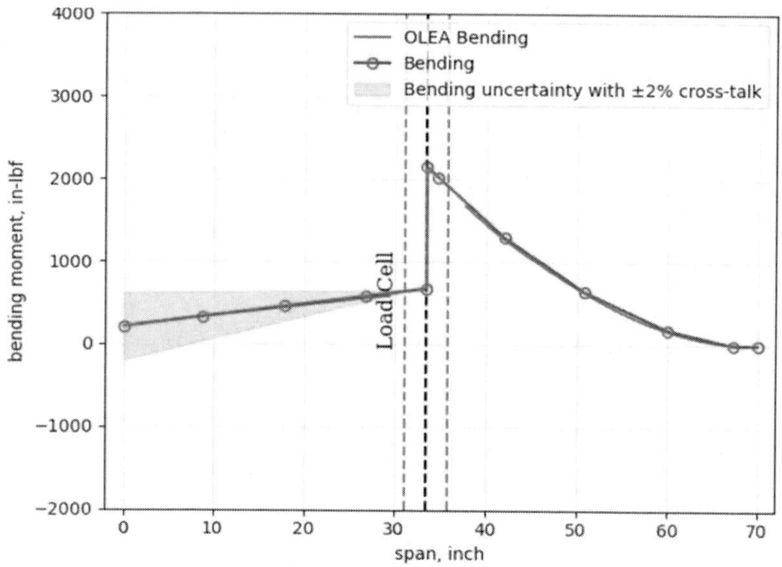

Figure B3. Out-of-plane bending moment diagram for wingtip-biased load case.

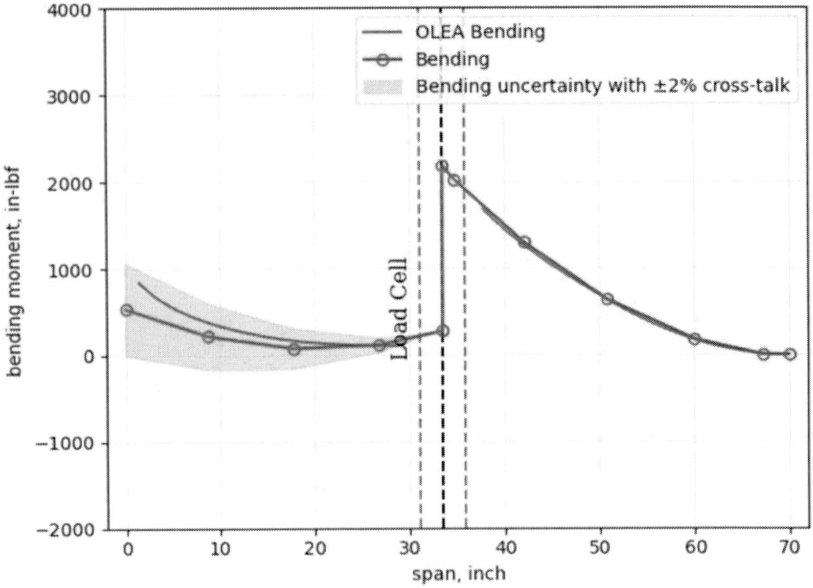

Figure B4. Out-of-plane bending moment diagram for uniform load case.

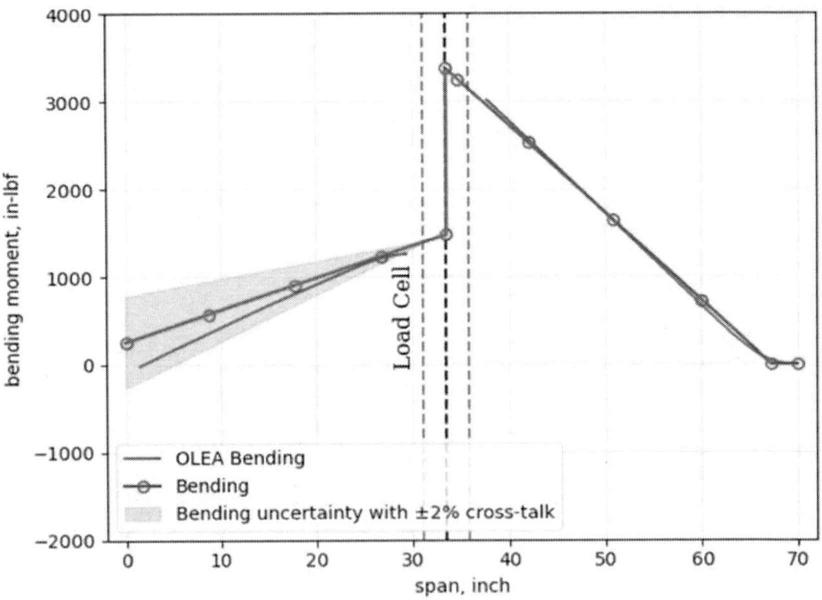

Figure B5. Operational Loads Estimation Algorithm out-of-plane bending moment diagram for concentrated wingtip load case.

Appendix C: Out-of-plane Shear Diagrams

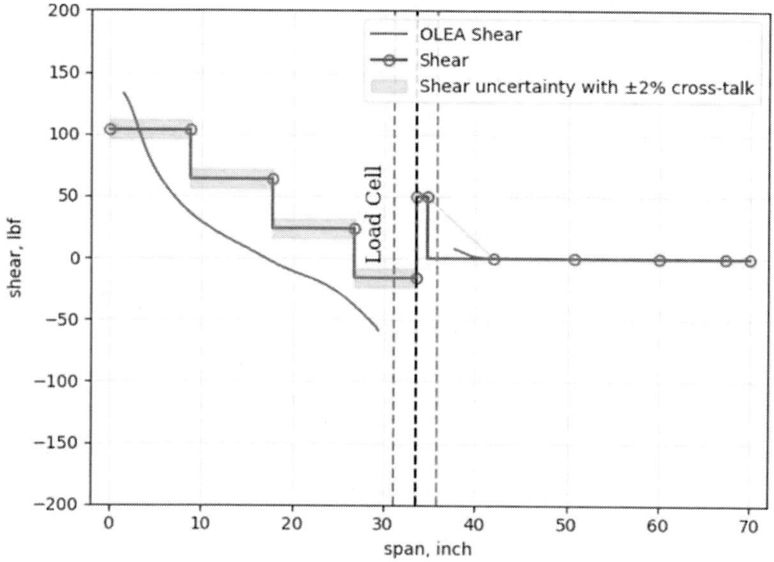

Figure C1. Out-of-plane shear diagram for root-biased load case.

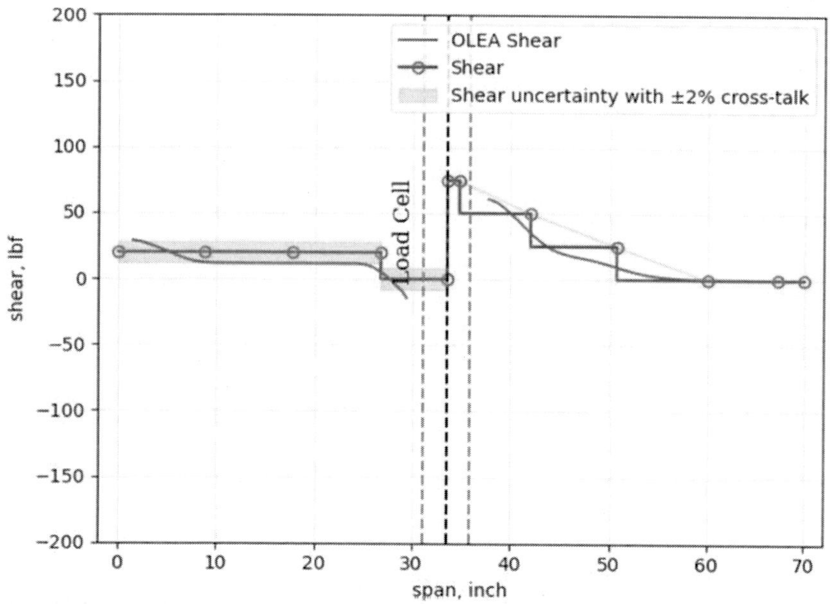

Figure C2. Out-of-plane shear diagram for mid-span-biased load case.

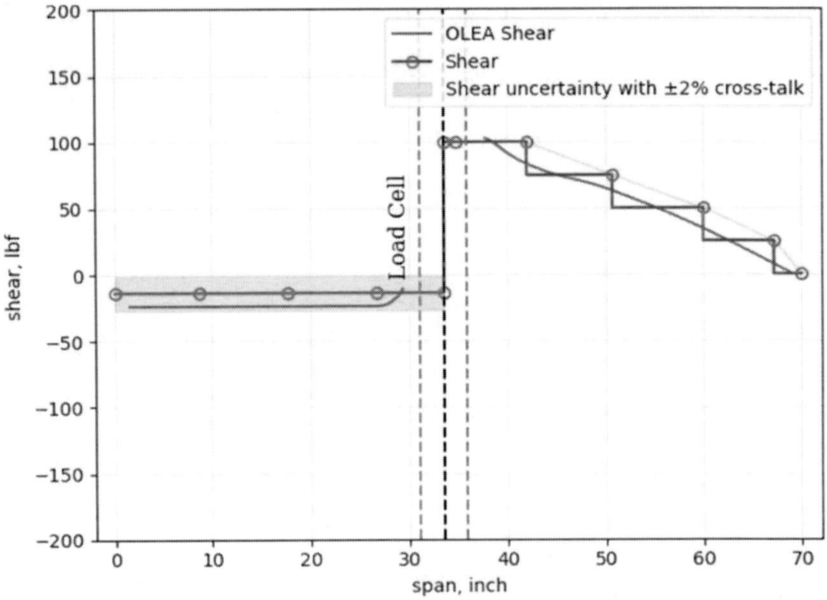

Figure C3. Out-of-plane shear diagram for wingtip-biased load case.

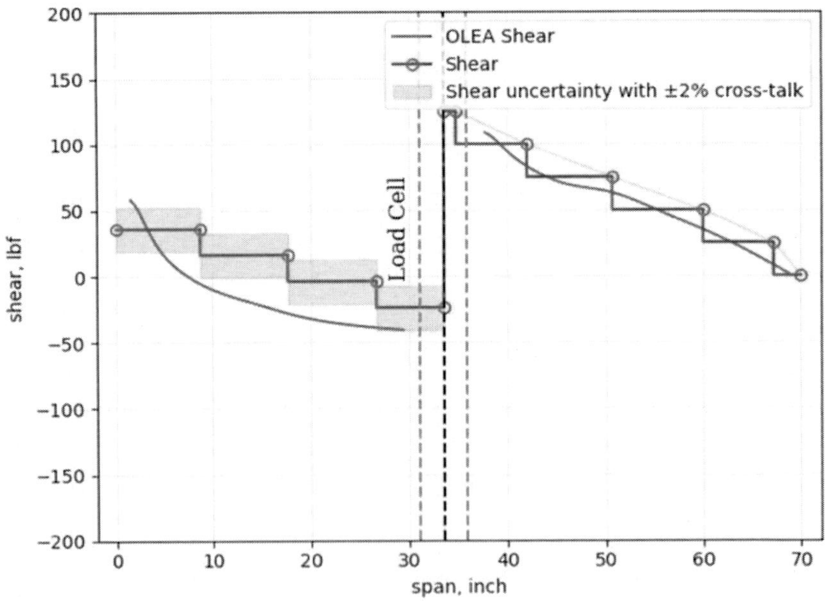

Figure C4. Out-of-plane shear diagram for uniform load case.

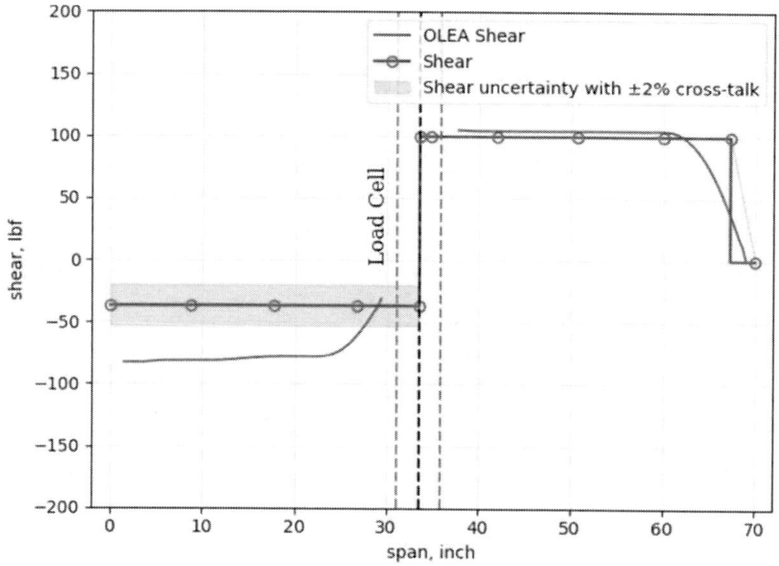

Figure C5. Out-of-plane shear diagram for concentrated wingtip load case.

Appendix D: In-plane Bending Diagrams

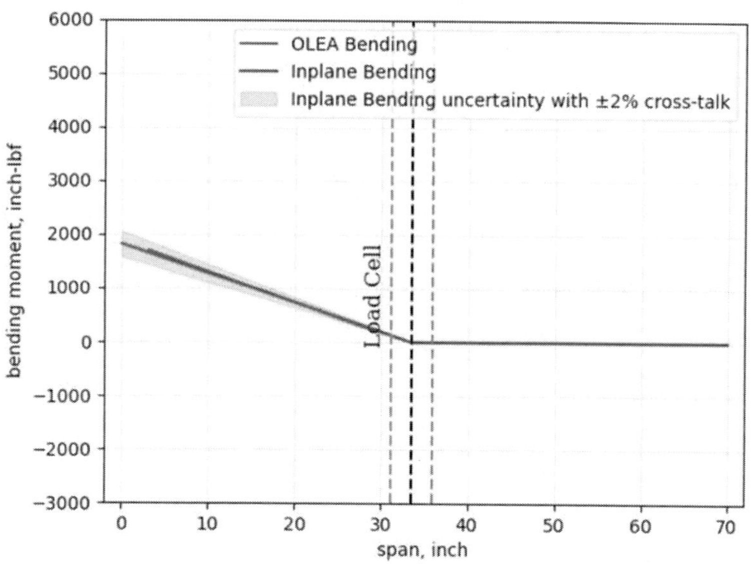

Figure D1. In-plane bending moment diagram for root-biased load case.

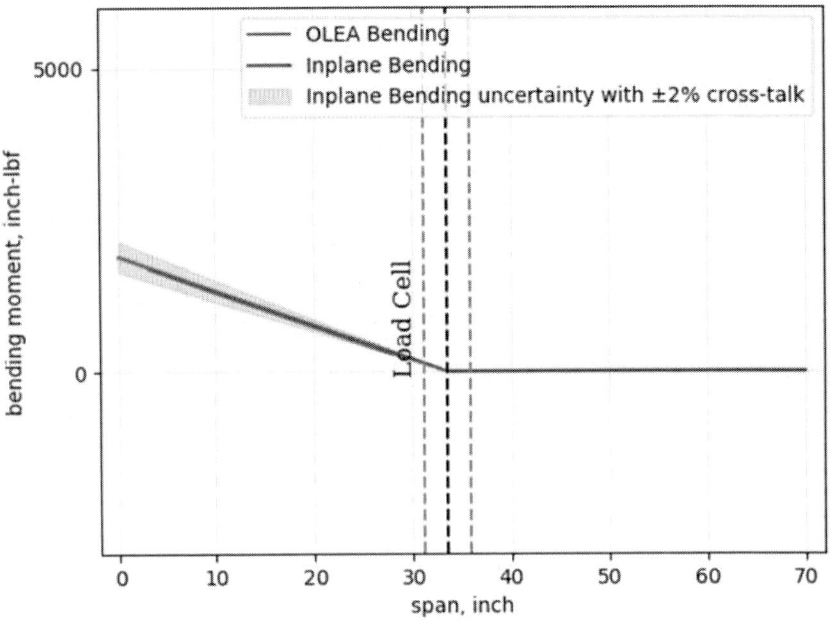

Figure D2. In-plane bending moment diagram for mid-span-biased load case.

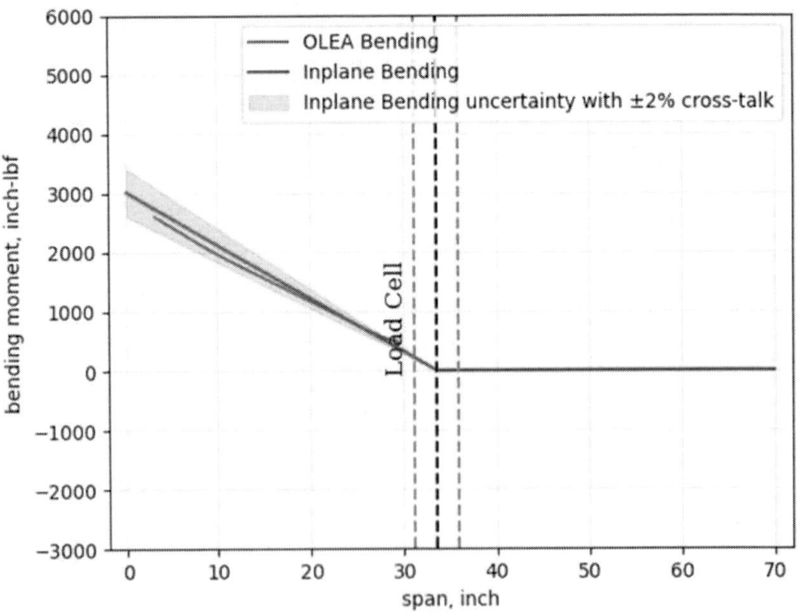

Figure D3. In-plane bending moment diagram for wingtip-biased load case.

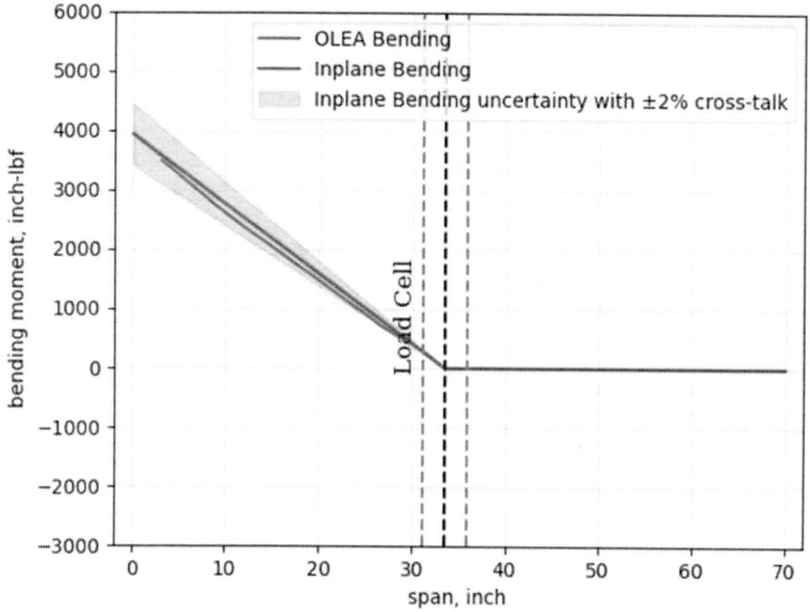

Figure D4. In-plane bending moment diagram for uniform load case.

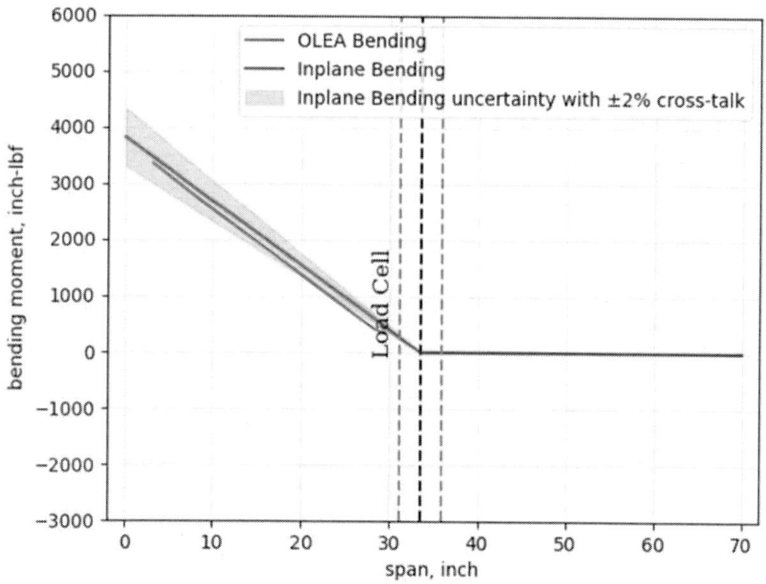

Figure D5. In-plane bending moment diagram for concentrated wingtip load case.

Appendix E: In-plane Shear Diagrams

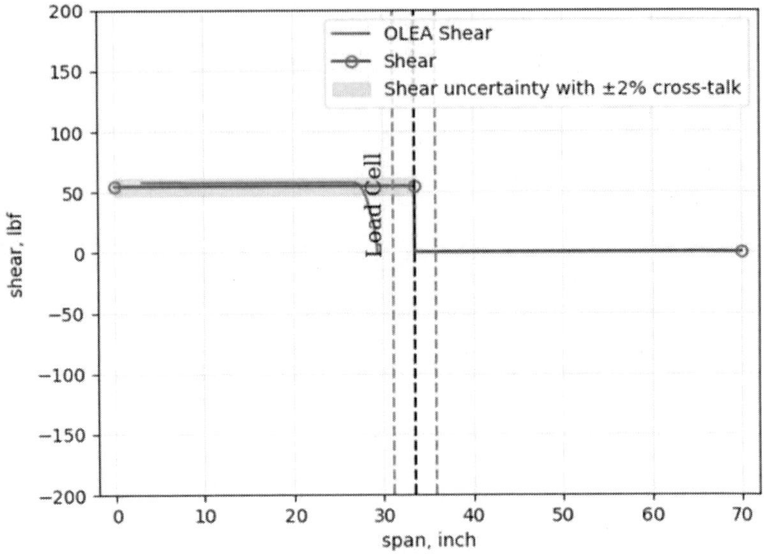

Figure E1. In-plane shear diagram for root-biased load case.

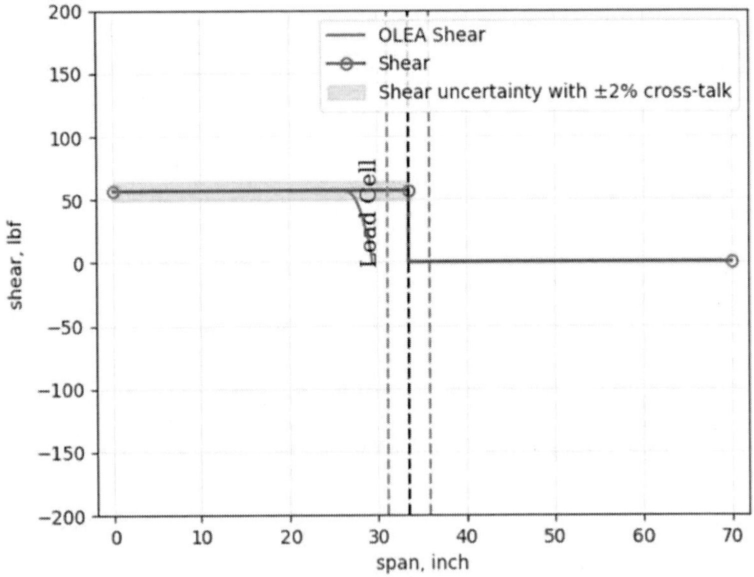

Figure E2. In-plane shear diagram for mid-span-biased load case.

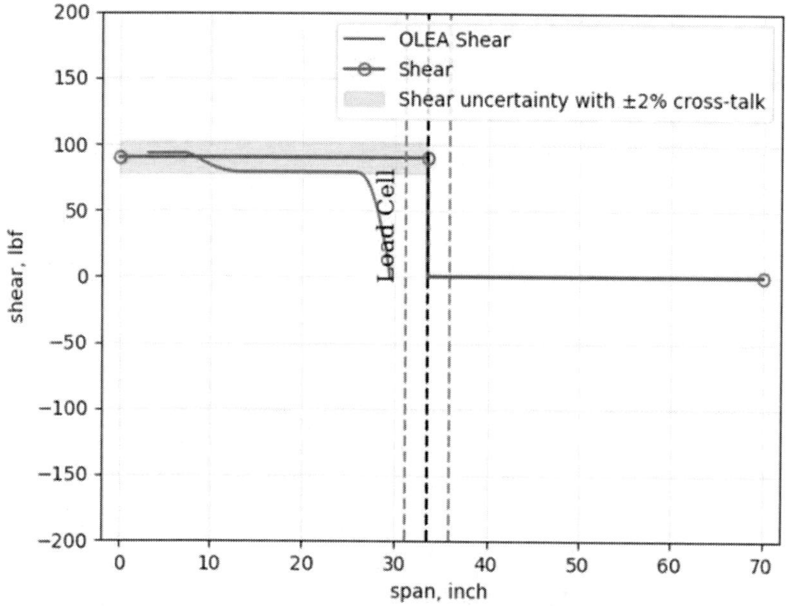

Figure E3. In-plane shear diagram for wingtip-biased load case.

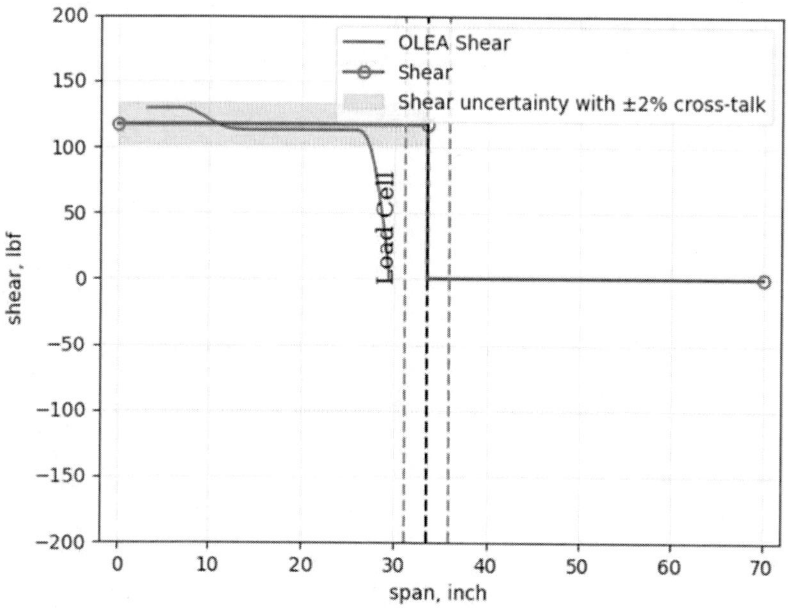

Figure E4. In-plane shear diagram for uniform load case.

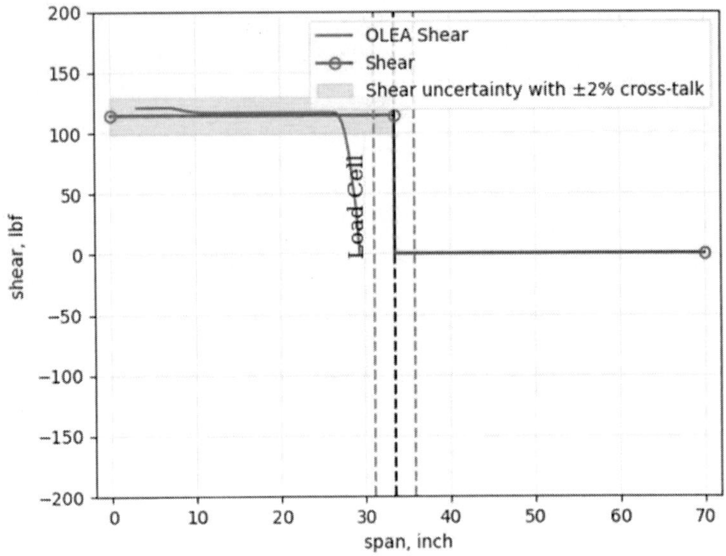

Figure E5. In-plane shear diagram for concentrated wingtip load case.

Appendix F: Out-of-Plane Displacement

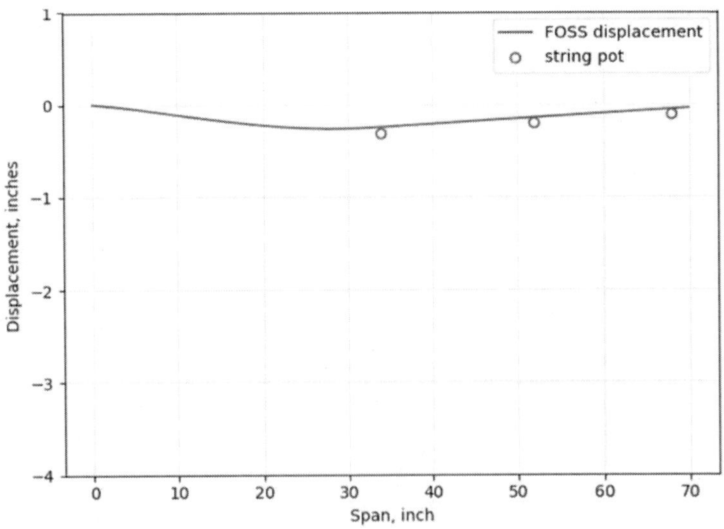

Figure F1. Out-of-plane displacement for root-biased load case.

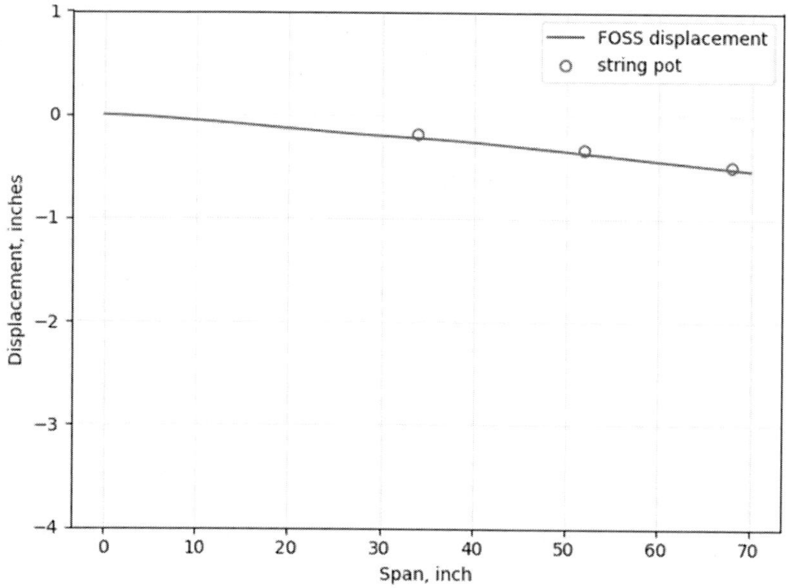

Figure F2. Out-of-plane displacement for mid-biased load case.

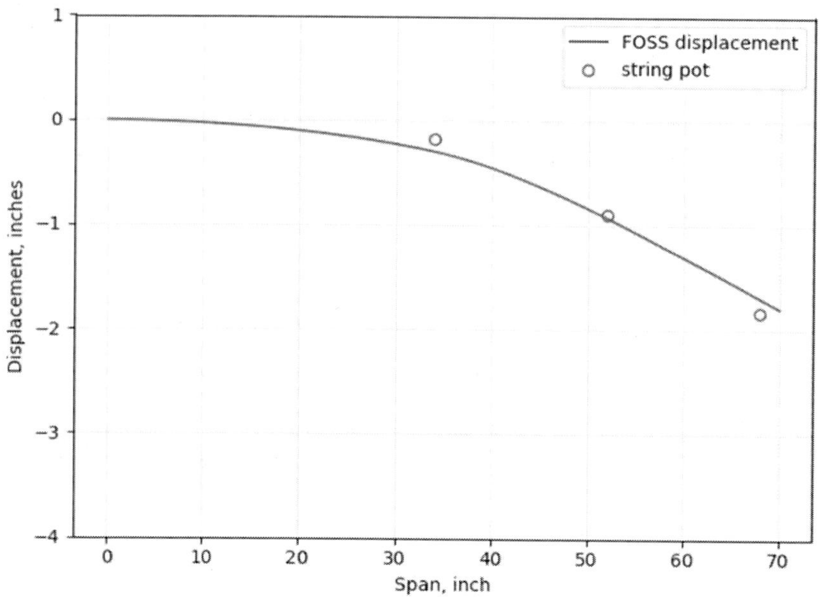

Figure F3. Out-of-plane displacement for tip-biased load case.

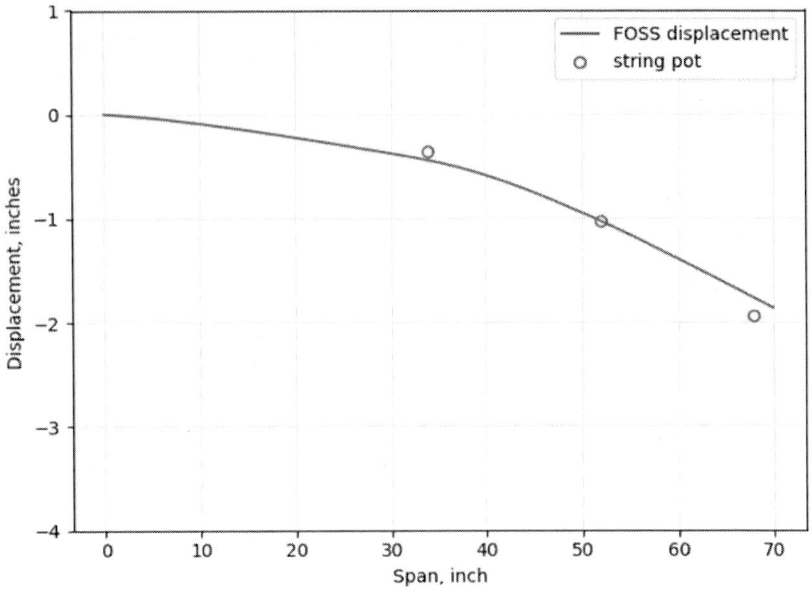

Figure F4. Out-of-plane displacement for uniform load case.

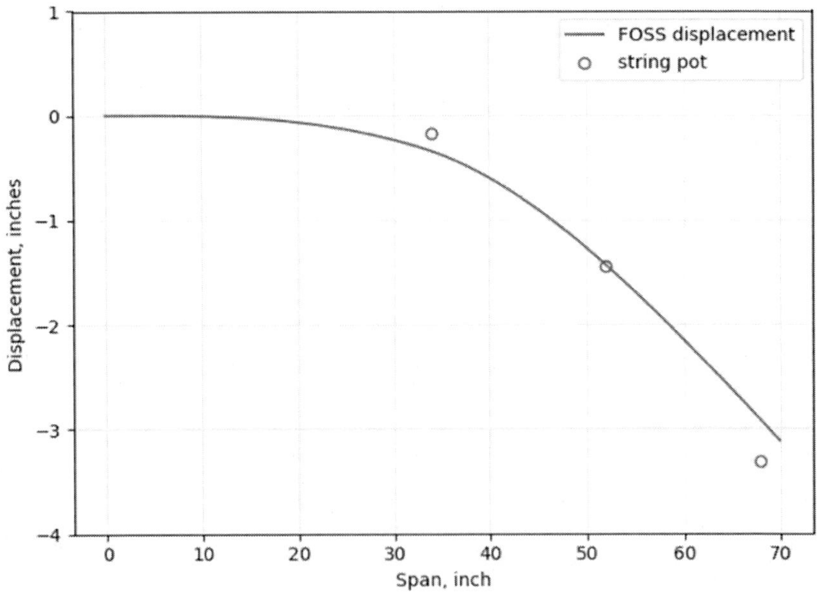

Figure F5. Out-of-plane displacement for concentrated wingtip load case.

References

[1] Viken, Sally A., Craig A. Hunter, S. Naomi McMillin, Gregory M. Gatlin, Daniel Maldonado, Jefferey A. Housman, Jared C. Duensing, James C. Jensen, and Cetin C. Kiris, *"Comparison of Computational Predictions of the Mach 0.80 Transonic Truss-Braced Wing Configuration with Experimental Data,"* NASA/TM-20205007879, June 2022.

[2] Skopinski, T. H., William S. Aiken, Jr., and Wilber B. Huston, *"Calibration of Strain-Gage Installations in Aircraft Structures for the Measurement of Flight Loads,"* NACA TR-1178, 1954.

[3] Richards, W. Lance, Allen R. Parker, Jr., William L. Ko, Anthony Piazza, and Patrick Chan, *"Application of Fiber Optic Instrumentation,"* NATO, RTO AGARDograph 160, Flight Test Instrumentation Series – Volume 22, SCI-228, 2012.

[4] Ko, William L., W. L. Richards, and Van T. Tran, *"Displacement Theories for In-Flight Deformed Shape Predictions of Aerospace Structures,"* NASA/TP-2007-214612, January 2007.

[5] Ko, William L., *"Further Development of Ko Displacement Theory for Deformed Shape Predictions of Nonuniform Aerospace Structures,"* NASA/TP-2007-214643,2007.

[6] Ko, William L., W. Lance Richards, and Van Tran Fleischer, *"Applications of Ko Displacement Theory to the Deformed Shape Predictions of the Doubly-Tapered Ikhana Wing,"* NASA/TP-2009-214652, 2009.

[7] Jutte, Christine V., William L. Ko, Craig A. Stephens, John A. Bakalyar, W. Lance Richards, and Allen R. Parker, *"Deformed Shape Calculation of a Full-Scale Wing Using Fiber Optic Strain Data from a Ground Loads Test,"* NASA/TP-2011-215975, 2011.

[8] Richards, W. L., and Ko, William L., *Process for Using Surface Strain Measurements to Obtain Operational Loads for Complex Structures*, United States Patent No. US 7,715,994, issued May 11, 2010.

[9] Ko, William L., and Van Tran Fleischer, *"Method for Estimating Operational Loads on Aerospace Structures Using Span-wisely Distributed Surface Strains,"* NASA/TP-2013- 216518, 2013.

[10] Bakalyar, J., and Jutte, C., "Validation Tests of Fiber Optic Strain-Based Operational Shape and Load Measurements," 53rd AIAA/ASME/ ASCE/AHS/ASC Structure, Structural Dynamics and Materials Conference, Honolulu, HI, AIAA Paper 2012-1904,, pp. 1-81.

[11] Budynas Richard. G., and Nisbett, J. Keith, *Shigley's Mechanical Engineering Design, Eighth Edition,* ISBN 978-0-07-312193-2, MHID 0-07-312193-2, McGraw-Hill, 2008.

[12] Pak, Chan-gi, "Unsteady Aerodynamic Force Sensing from Strain Data," *Journal of Aircraft, Vol.* 54, No. 4, July 2017.

Chapter 3

SCIFLI HORIS Windowpane Optical Analysis*

Hyun Jung Kim
William Humphreys
Arianna Haven
Carey Scott
Ian Rook
Thomas Gutierrez
Alexandra Philip
and Tait Pottebaum

NASA STI Program Report Series

Since its founding, NASA has been dedicated to the advancement of aeronautics and space science. The NASA scientific and technical information (STI) program plays a key part in helping NASA maintain this important role.

The NASA STI program operates under the auspices of the Agency Chief Information Officer. It collects, organizes, provides for archiving, and disseminates NASA's STI. The NASA STI program provides access to the NTRS Registered and its public interface, the NASA Technical Reports Server, thus providing one of the largest collections of aeronautical and space science STI in the world. Results are published in both non-NASA channels and by NASA in the NASA STI Report Series, which includes the following report types:

---

* This is an edited, reformatted and augmented version of NASA Publication No. NASA/TM–20240007129, dated July 2024.

In: Applications of Optical Measurement Techniques …
Editor: Octavian Melina
ISBN: 979-8-90134-275-6

- x TECHNICAL PUBLICATION. Reports of completed research or a major significant phase of research that present the results of NASA Programs and include extensive data or theoretical analysis. Includes compilations of significant scientific and technical data and information deemed to be of continuing reference value. NASA counterpart of peer-reviewed formal professional papers but has less stringent limitations on manuscript length and extent of graphic presentations.

- x TECHNICAL MEMORANDUM. Scientific and technical findings that are preliminary or of specialized interest, e.g., quick release reports, working papers, and bibliographies that contain minimal annotation. Does not contain extensive analysis.

- x CONTRACTOR REPORT. Scientific and technical findings by NASA-sponsored contractors and grantees.

- x CONFERENCE PUBLICATION
Collected papers from scientific and technical conferences, symposia, seminars, or other meetings sponsored or co-sponsored by NASA.

- x SPECIAL PUBLICATION. Scientific, technical, or historical information from NASA programs, projects, and missions, often concerned with subjects having substantial public interest.

- x TECHNICAL TRANSLATION. English-language translations of foreign scientific and technical material pertinent to NASA's mission.

Specialized services also include organizing and publishing research results, distributing specialized research announcements and feeds, providing information desk and personal search support, and enabling data exchange services.

For more information about the NASA STI program, see the following:

- x Access the NASA STI program home page at http://www.sti.nasa.gov

Help desk contact information: https://www.sti.nasa.gov/sti-contact-form/ and select the "General" help request type.

Abbraviations

AR	anti-reflective
ASTM	American Society for Testing and Materials
BET	best estimated trajectory
CA	Clear Aperture
CFD	Computational Fluid Dynamics
ED	Engineering Directorate
EEV	Earth entry vehicle
FS	fused silica
HORIS	Hypervelocity OSIRIS-REx Reentry Imaging and Spectroscopy
ISO	International Organization for Standardization
JSC	Johnson Space Center
KSC	John F. Kennedy Space Center
LDTD	Laboratories, Development and Testing Division
LaRC	Langley Research Center
MTF	modulation transfer function
OSIRIS-Rex	Origins, Spectral Interpretation, Resource Identification and Security Regolith Explorer
PSF	point spread function
PV	peak to valley
RMS	root-mean squared
SAMI	SCIFLI Airborne Multispectral Imager
SCIFLI	Scientifically Calibrated In-Flight Imagery
SN, S/N	serial number
SNR	signal to noise ratio
SRC	sample return capsule
TPS	Thermal Protection System
TWE	transmitted wavefront error
UV	ultraviolet
VIS	visible
WFE	wave front error

Nomenclature

$E(\lambda)$	White light source characteristics
k	Normalization factor for tristimulus calculation

$R(\lambda)$	Optical reflectance
$T(\lambda)$	Optical transmittance
$T1$	Haze measurement incident light
$T2$	Haze measurement total transmitted light
$T3$	Haze measurement scattered light by instrument
$T4$	Haze measurement scattered light by instrument and specimen
x, y	Chromaticity coordinates
$\bar{x}(\lambda), \bar{y}(\lambda), \bar{z}(\lambda)$	*Standard observer characteristics*
X, Y, Z	Tristimulus values
λ	Wavelength (nm)

Executive Summary

This NASA Technical Memorandum summarizes windowpane optical analyses for the OSIRIS-REx (Origins, Spectral Interpretation, Resource Identification and Security Regolith Explorer) mission [1] supported by the SCIFLI (Scientifically Calibrated In-Flight Imagery) team [2] at NASA Langley Research Center (NASA LaRC). The SCIFLI team led a NASA multi-center team in characterization of fused silica (FS) and anti-reflective coated (AR-coated) sapphire materials for use as aircraft windowpanes. These materials were needed for the Hypervelocity OSIRIS-REx Reentry Imaging & Spectroscopy (HORIS) mission and were tested to generate performance parameters relevant to the flight environment. Some of the assessed parameters included optical quality, transmittance, reflectance, color balance, birefringence, transmitted wavefront error, wedge, and haziness of the windows.

In order to evaluate the material characteristics in a uniform manner, several criteria were held constant. For example, the physical dimensions of the test articles were held constant (i.e., 17-inch diameter aircraft windowpane and 2-inch diameter witness samples). In this way, the comparison of test results from one material specimen to another was invariant with regard to the manner in which the tests were conducted. This also reduced the variability in how the data were reported. As a result, the results shown in this report, including test data generated by NASA can be used for windowpane material evaluation and future material design trades. Window designers can use the data and the testing method along with the testing matrices to efficiently match material performance to their design and mission objectives and requirements.

Standardized testing methodologies and data reduction procedures are described in this report. Unless otherwise noted in the data itself, all the windowpane materials were tested in order to characterize their performance for the HORIS mission.

One appendix has been included in this report. These include detailed wavefront interferograms of transmitted wavefront error and optical wedge used in the generation of aggregated material performance.

Background

NASA's OSIRIS-REx (Origins, Spectral Interpretation, Resource Identification and Security Regolith Explorer) is the first U.S. mission to collect an asteroid sample and deliver it to Earth. Asteroid Bennu was selected for this mission for several reasons. Chief among them was its proximity to Earth and its size, both of which made it easier for a spacecraft to reach, orbit, and collect a sample from the asteroid. Bennu's carbon-rich composition, deduced by studying its spectral properties from Earth, also made it an attractive study target, particularly as scientists search for chemical markers that might have played a role in life's origins. In addition, scientists will be able to use data about Bennu's orbit to further study the Yarkovsky Effect, a force acting on a rotating body in space caused by the anisotropic emission of thermal photons, which carry momentum [1], or how sunlight and heat affect an asteroid's path, a critical factor in planetary defense.

OSIRIS-REx launched September 8, 2016, aboard an Altas V rocket from NASA's Kennedy Space Center, then reached Bennu in 2018. It spent two years in orbit as it mapped the asteroid's surface to help the mission team find the optimal spot for sample collection. Bennu proved to be rockier and rougher than the mission was designed for, making sample collection a much bigger challenge. Nevertheless, although two sites were selected, sampling the first, nicknamed Nightingale, was so successful that a second attempt was unnecessary. On October 20, 2020, the spacecraft successfully collected an estimated 8.8 ounces of material from the asteroid. After orbiting the Sun twice, OSIRIS-REx returned to Earth on September 24, 2023. Upon return, the capsule containing the Bennu sample separated from the rest of the spacecraft and entered Earth's atmosphere. The capsule parachuted to the Utah Test and Training Range (Dugway) where scientists retrieved it [3]. An extended mission will take the OSIRIS-REx spacecraft into orbit around the near-Earth asteroid Apophis in 2029 [4].

HORIS Mission

The Scientifically Calibrated In-Flight Imagery (SCIFLI) team at the NASA Langley Research Center collects real -time visual, infrared, and spectral data on vehicles while in flight. Doing so helps researchers gain insight into some of the most challenging questions in fluid dynamics, provides essential engineering data, and documents vital flight safety systems. With each successful mission [2], SCIFLI extends the experimental nature of wind-tunnel testing and computational aerothermal analysis by observing vehicles under real flight conditions. In addition to collecting these invaluable data, SCIFLI supports commercial and NASA missions by ensuring the safety of the vehicles and astronauts aboard. The topic of this report is a recent SCIFLI observation campaign supporting OSIRIS-REx, namely the Hypervelocity OSIRIS-REx Reentry Imaging and Spectroscopy (HORIS) mission. The HORIS mission included three ground imaging sites along with the four aircraft as shown in Figure 1.

The measurements and analyses goals for the observation campaign include the following:

1) Near-vehicle wake radiation measurements
   - UV–VIS (315–700 nm): N2, NO, and NO2 band systems
   - Plasma species radiative heating to leeward (payload) side of reentry vehicle
2) Broadband, low-noise shock-layer measurements
   - NIR–SWIR (700–2100 nm): Shock heated air species N, O, N2, N2+
   - Nonequilibrium electron number density: Hβ VIS (486 nm)
   - Computational fluid dynamics (CFD) model validation and tool development for other Earth entry vehicles (EEVs)
3) Thermal Protection System (TPS) surface temperature analysis
   - VIS–NIR (400–950 nm): Broadband thermal emission from surface
   - TPS material response to allow validation of NASA CFD tools.

4) Trajectory reconstruction for best estimated trajectory (BET) derivation
    - NIR–SWIR (700–1800 nm)
    -
    - Fielding multiple systems.

Overall mission goals:

1) Thermal emissions from the sample return capsule (SRC) surface and shock-layer during peak heat loading
2) Plasma species relaxation-related radiative heating to the leeward (payload) side of the reentry vehicle
3) Shock-heated air species and non-equilibrium electron number densities
4) High signal-to-noise ratio (SNR) data during a daylight entry event

These datasets were designed to help researchers better understand high-speed atmospheric entries and how the OSIRIS-REx capsule's thermal protection system functioned under extreme conditions. Given the stringent science goals for the mission, it was critical that the highest quality windows be installed on aircraft to permit the acquisition of imagery data in the desired optical spectral wavebands. Thus, a laboratory investigation was conducted to characterize the performance of various window candidates to achieve the HORIS mission objectives.

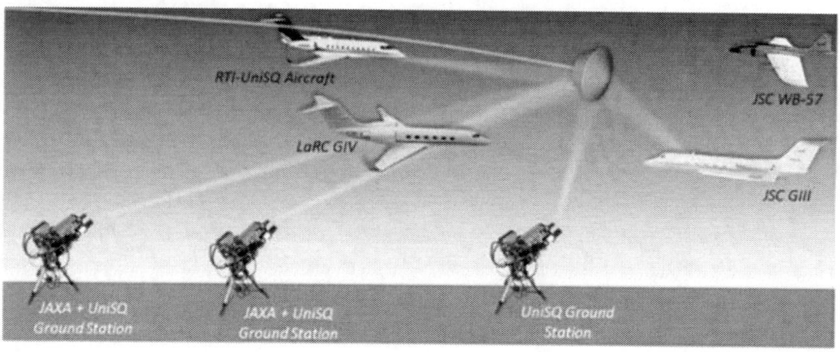

Figure 1. Four imaging aircraft and three ground imaging sites for science goals and imaging objectives of the HORIS mission.

Figure 2 shows examples of airborne imaging windows installed on Gulfstream aircraft stationed at NASA LaRC (tail number NASA 522) and

Johnson Space Center (JSC) (tail number NASA 992), with the corresponding window frame assemblies shown in Figure 3. These assemblies incorporate a 17-inch diameter aperture and utilize single-pane optical windows. The choice of window enhances the quality of data collected in airborne imaging missions by utilizing a single pane element made of a highly transmissive material in the wavebands of interest (i.e., ultraviolet, visible, and midwave infrared). Due to the very long optical paths employed in the measurements (50–200 km away from the imaging target's trajectory), any improvement to window performance permits the acquisition of higher-quality data sets.

Figure 2. Fused Silica window utilizing a 17-inch diameter (16.4-inch diameter viewable area) with custom frames compatible with Gulfstream window ports and installed on a Gulfstream 4 Aircraft at NASA LaRC: (a) outside view, (b) inside view, and (c) in-flight view during a mission.

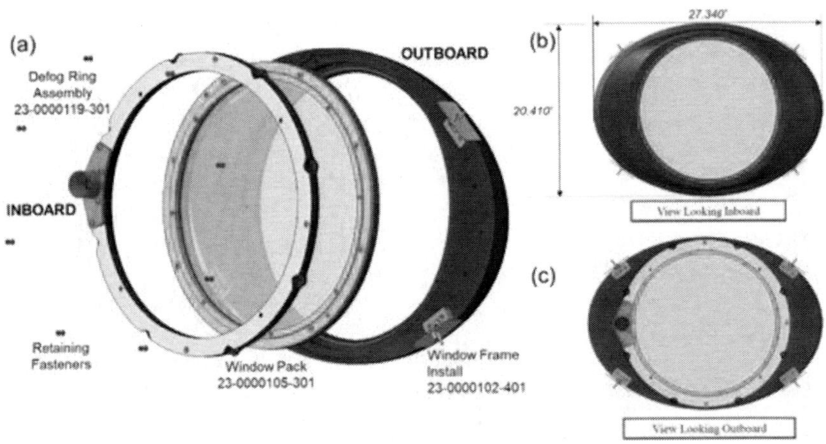

Figure 3. Optical window frames: (a) window frame assembly; outer frame (dark gray) is potted into the aircraft frame, and the window pack (green and red) is held into the outer frame with the defog ring (yellow and blue) and eight fasteners, (b) view looking inboard with dimensions, and (c) view looking outboard.

Figure 4. External view of windows on NASA Gulfstream aircraft, left to right: original gulfstream window, previously designed acrylic "racetrack" shaped window, two new fused silica windows, and an original Gulfstream window.

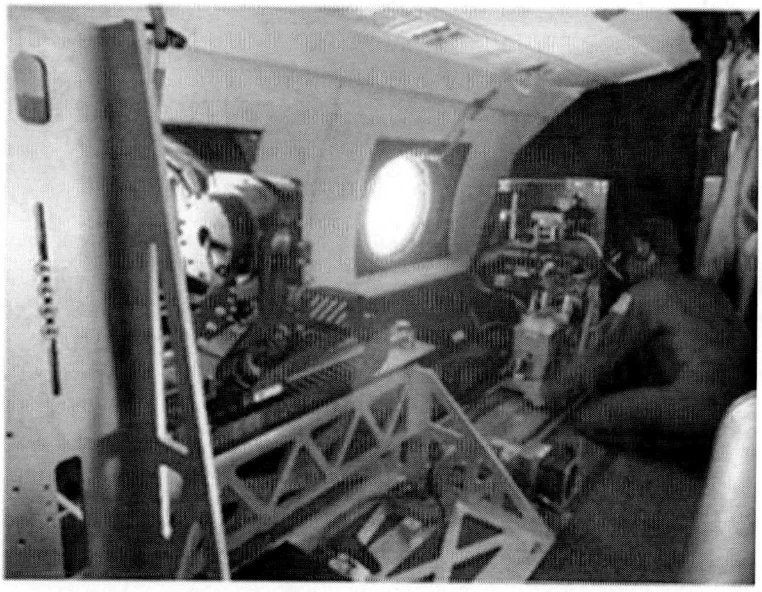

Figure 5. Imaging systems installed in front of fused silica windows (in-flight).

In early 2023, seven new, single-pane optical quality windows (six fused silica and one sapphire) were designed and acquired by SCIFLI specifically for the HORIS mission, where airborne imagery data were acquired during the reentry of the SRC as it landed in the Utah desert. The windows were designed to be compatible with the NASA Gulfstream aircraft, with installation in selected window ports as shown in Figure 4. The two NASA Gulfstream aircraft utilized airborne imaging systems similar to those shown in Figure 5.

Figure 6 shows 17-inch fused silica (FS) window pairing on LaRC GIV (N522NA) and JSC GII (N922NA). Each fused silica window is assigned a sixteen-digit serial number, marked on the edge of the window in indelible ink. The final five numbers, which will be used throughout this paper, are defined as 9yzz, where y =0 for an uncoated mirror and y =1 for a coated mirror and zz is a sequence number starting with the number 10. The six fused silica mirrors are numbered 9010 through 9015.

Optical Metrology Overview

Optical metrology refers to a suite of measurements that use light to gather data regarding the physical properties of materials (or objects). The measurement techniques use light to interact with the material under study, and by measuring the characteristics of the light that is reflected, refracted, or absorbed by the material the optical properties of the material can be extracted. Typical measurands obtained for transparent materials include (1) transmittance, (2) reflectance, (3) color balance, (4) birefringence, (5) transmitted wavefront, (6) optical wedge, and (7) optical haze. These measurements are described in this section.

Transmittance, Reflectance and Color Balance: The purpose of spectral transmittance testing is to measure the optical transmittance as a function of optical wavelength across the visible and/or NIR spectral ranges, depending on the material under inspection [9-13]. The measurement is performed using a commercial spectrometer and spectrophotometer. The optical transmittance is the percentage ratio between the incoming and outgoing intensities of an optical beam as it propagates normally through a transparent specimen. The spectral transmittance of a material provides three important pieces of information:

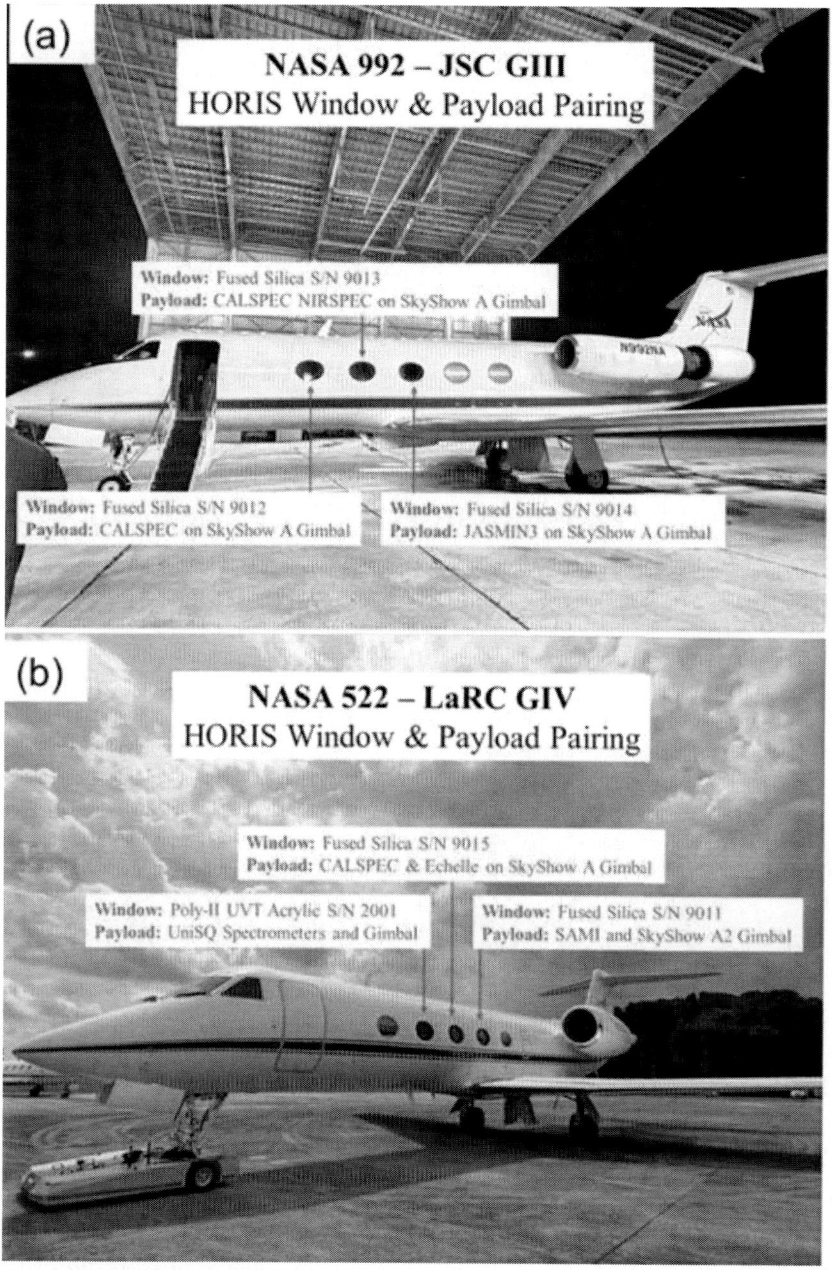

Figure 6. Payload and Window Pairing on (a) JSC GIII (N922NA) aircraft and (b) LaRC GIV (N522NA) aircraft.

1) It provides a direct measure of the degree of attenuation that the material will introduce the incoming optical intensity at every spectral component.
2) It provides the spectral reflectance, obtained computationally since normal reflectance is very difficult, if not impossible, to obtain using direct measures.
3) It provides a measure of the color balance, which is a quantity of how far a defined white input light will deviate in color as it propagates across the material.

Birefringence: The objective of optical birefringence testing is to measure the difference between two refractive indices along two orthogonal directions of a linearly polarized optical beam per unit length of the specimen thickness; i.e., the optical retardance per unit thickness of the specimen in centimeters (cm).

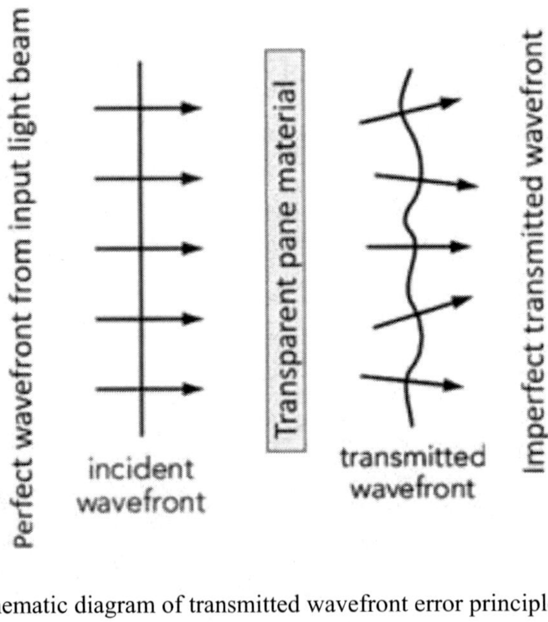

Figure 7. Schematic diagram of transmitted wavefront error principle.

Transmitted Wavefront Error (TWE): The purpose of transmitted wavefront error (TWE) testing is to determine the change in wavefront of a defined input light beam as it passes through the specimen under inspection. The error in the transmitted wavefront provides important information on the optical performance of the pane material surfaces (see Figure 7). The test yields data for peak to valley (PV) TWE with piston, tilt, and defocus parameters removed and with only piston and tilt parameters removed. It also measures root-mean squared (RMS) TWE with piston, tilt, and defocus parameters removed and with only piston and tilt parameters removed. Using a Zygo (Middlefield, Connecticut) interferometer with a 6-inch size aperture, the point spread function (PSF) in addition to the modulation transfer function (MTF) of the material can be acquired. Note that the full optical transfer function is the Fourier transform of the PSF, and if needed, the phase information can be obtained as well.

Optical Wedge: Optical wedge is a measure of the parallelism between the two surfaces of an optical specimen within a defined viewing area. Typically, the wedge magnitude is obtained between the front and back surfaces of the specimen; i.e., this quantifies the net deviation of the specimen from parallelism. A Fizeau interferometer is the common instrument for measuring optical wedges.

Optical Haze and Clarity [10]: Optical haze measurements quantify the percentage of transmitted light that deviates from an incident beam direction by more than 2.5 degrees from the surface normal. The same measurement system that is used to measure haze can also measure optical clarity and luminous transmittance. Optical clarity is the percentage of transmitted light that is evaluated at angles of less than 2.5 degrees, and luminous transmittance is the percentage of the transmitted light that can be sensed by human vision, weighted by the human eye sensitivity curve. Commercial instruments are typically used for haze and clarity non-contact measurements.

Window Specifications

Figures 8 and 9 depict SCIFLI 17-inch diameter fused silica and anti-reflective- (AR-) coated sapphire optical window specifications. All the optical properties (transmittance, reflectance, color balance, birefringence, wedge, and haze) that could meet the mission requirements were specified. For scientific grade windows, typically the transmitted wave front error (WFE) (the amount of distortion of the wavefront through the window) should ideally be λ (lambda)/10 or less. Note that 1/10 waves are considered a standard for high-quality optics [11], [12]. Maintaining this requirement over the full 16.4 clear aperture by the vendor was challenging and the windows needed to be properly polished and/or coated to meet the performance specification. For the HORIS mission, NASA accepted a relaxed specification, better than λ/8, depending on how the measurements were obtained and handled. Specifically, NASA stated that for normal incidence viewing with piston, tilt, and defocus removed (the three most predominant aberrations), the RMS wavefront error over the window should not exceed λ/8 wave. This was a relaxed requirement versus the standard λ/10 standard; however, it was agreed upon since the mission was using an acrylic window as well.

Items Desired:	Qty(6) 17-inch dia. Optical Grade Fused Silica Window
Material:	Corning 7980 Fused Silica
Shape:	Round, cylinder
Diameter:	D = 17.000" +0.000/-0.040
Clear Aperture:	16.40 inches +/- 0.01
Thickness:	1.300 +/- 0.005
Temperature Range:	-70°F / +150°F
Surface Finish:	All non-polished surfaces very fine-ground; ≥ 220 Grit Finish (0.5 – 1.0 Ra[micron])
Edge:	- All edges 0.05 +/- 0.01 X 45° chamfer (see Detail A on drawing *23-0000112.pdf*) - Edge chips shall conform to MIL-PRF-13830. Paragraph 3.7.9.2 - Maximum radius of 0.005" on edges, no sharp corners - Bevels may be polished to achieve wavefront requirements. 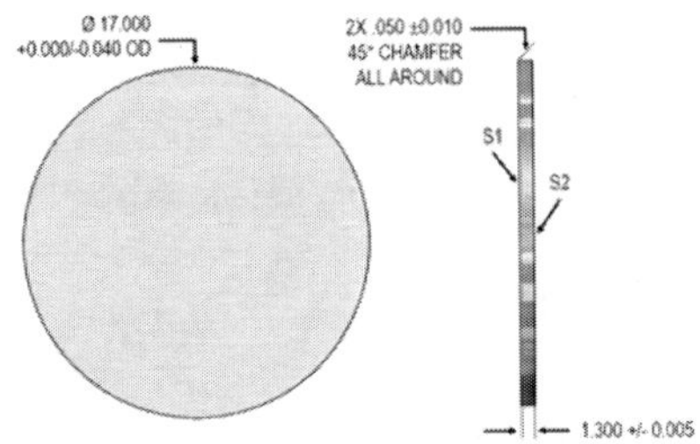
Parallelism:	The two main surfaces of the window, "S1" and "S2," shall be parallel within 0.002 inches
Transmitted WFE:	- The transmitted wavefront error shall be measured in collimated light at 632.8 nm with light transmitted over the full clear apertures designated on sides "S1" and "S2". (Lacking a full aperture interferometer, the interferograms can be stitched together. Minimum sub-aperture size of 6" diameter) - λ/10 waves peak to valley @ 632.8 nm at normal incidence across entire clear aperture
Surface Quality:	- Scratch/Dig = 60/40 on both sides "S1" and "S2" - Inspection report including map (preferably on transparent overlay) of scratches and digs to be provided with each window
Notes:	- Cracks are not acceptable anywhere on or in the window. - Qty(6) 2-inch diameter witness samples from same stock and thickness and finished to same quality shall be provided along with the window. - Optical characterization and transmission chart of the window provided with window delivery. Any Exceptions shall be discussed with customer

Figure 8. FS optical window specification.

Items Desired:	Qty(1) 17-inch dia. Optical Grade Sapphire Window
Material:	Sapphire
Orientation:	C-plane, +/- 0.05 deg
Shape:	Round, cylinder
Diameter:	D = 17.000" +0.000/-0.040
Clear Aperture:	16.40 inches +/- 0.01
Thickness:	0.750 +/- 0.005
Temperature Range:	-70°F / +150°F
Surface Finish:	All non-polished surfaces very fine-ground; $\geq$ 220 Grit Finish (0.5 – 1.0 Ra[micron])
Edge:	- All edges 0.05 +/- 0.01 X 45° chamfer (see Detail A on drawing *23-0000106.pdf*) - Edge chips shall conform to MIL-PRF-13830. Paragraph 3.7.9.2 - Maximum radius of 0.005" on edges, no sharp corners - Bevels may be polished to achieve wavefront requirements. 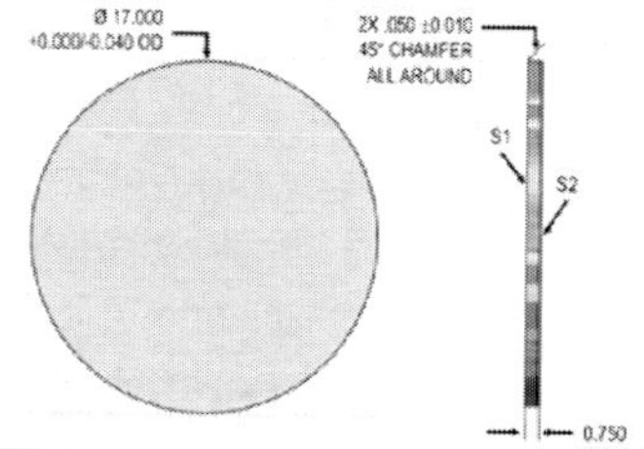
Parallelism:	The two main surfaces of the window, "S1" and "S2," shall be parallel within 0.002 inches
Transmitted WFE:	- The transmitted wavefront error shall be measured in collimated light at 632.8 nm with light transmitted over the full clear apertures designated on sides "S1" and "S2". (Lacking a full aperture interferometer, the interferograms can be stitched together. Minimum sub-aperture size of 6" diameter) - $\lambda/10$ waves peak to valley @ 632.8 nm at normal incidence across entire clear aperture
Surface Quality:	- Scratch/Dig = 60/40 post coating on both sides "S1" and "S2" - No visible inclusions, no visible scratches
Coating:	- Surfaces "S1" and "S2" shall have a hard dielectric antireflective coating applied over the full aperture. AR coating shall meet or exceed the following transmission requirements over the clear aperture: **BBAR coating from 0.31 – 5 µm with following:** ○ 0.31 to 0.6 um Rmax less than 3.5% per surface ○ 0.62 to 0.75 um Rmax less than 5.5%, Ravg less than 4.0% per surface ○ 0.8 to 0.95 um Rmax less than 6.0%, Ravg less than 5.0% per surface ○ 0.988 to 1.1 um Rmax less than 5.5%, Ravg less than 5.0% per surface ○ 1.28 to 1.35 um Rmax less than 5.0%, Ravg less than 4.0% per surface ○ 1.58 to 1.65 um Rmax less than 4.5%, Ravg less than 3.75% per surface ○ 2.18 to 2.3 um Rmax less than 5.5%, Ravg less than 4.75% per surface ○ 3.58 to 4.1 um Rmax less than 6.5%, Ravg less than 6.0% per surface ○ 4.58 to 5.0 um Rmax less than 6.5%, Ravg less than 6.0% per surface ○ Durability will be verified on simultaneously coated witness sample per MIL-PRF-13830B, C.3.8.2 (humidity), C.3.8.4.2 (moderate abrasion) and C.3.8.5 (adhesion). ○ All untoleranced surfaces: 200 grit or finer surface finish ○ Cosmetic quality, including surface quality, will be verified using MIL-PRF-13830B ○ Vendor will provide interferometric data that indicates conformance to the optical specifications. ○ Vendor will provide measured reflectivity of window surfaces or witness samples. - The coating specs should also include the range of angles of incidence, which is approximately -15° to 45°. A narrower nominal range of incident angles may be defined. - Anti-reflective coating applied to surfaces "S1" and "S2" may extend beyond the clear aperture. - Coatings shall meet the following requirements of MIL-PRF-13830B Appendix C Section 3.8 for: 1. Adhesion 2. Temperature 3. Severe abrasion 4. Salt solubility 5. Water solubility - All Specifications apply over the clear aperture only.
Notes:	- Cracks are not acceptable anywhere on or in the window. - Qty(10) 2-inch diameter **witness samples** shall be provided along with the window - Optical characterization and transmission chart of the window provided with window delivery. - Any Exceptions shall be discussed with customer

Figure 9. Sapphire optical window specifications.

HORIS Window Optical Metrology

HORIS candidate window optical characterizations were conducted at Kennedy Space Center (KSC) in the Metrology and Precision Measurement Alignment Lab. The SCIFLI team submitted six 17-inch diameter FS windows and six 2-inch diameter FS witness samples for testing. The windowpanes were categorized as Category A based on requirements defined in Ref. 14. Category A windows (Custom Optical Window Ports) are window-based customer-unique requirements developed on a case-by-case basis. The SCIFLI team compiled custom requirements for the windows and this information was relayed to the KSC Metrology Lab. The metrology testing matrix is summarized in Table 1. This section discusses specific testing methodologies that were used in characterizing the window performance.

Table 1. Window testing matrix for HORIS mission

Samples	Quantities Total / Pane Serial #	Wavelength	Transmittance, Reflectance and Color Balance Testing (Y/N)	Birefringence Testing (Y/N)	Transmitted Wavefront Error (TWE) Testing (Y/N)	Wedge Testing (Y/N)	Haze Testing (Y/N)
17 " dia. Fused silica (quartz) ‡	6 23-0000112-001-9010, 9011, 9012, 2013, 9014, 9015	320 nm-800 nm in 5 nm step sizes	Y	Y†	Y	Y	N
2" dia. Fused silica (quartz)	6 SN01, 03, 04, 05, 06, 07	180 nm-3000 nm in 5 nm step sizes	Y*	N	N	N	Y
17" dia. AR-coated sapphire	1	320 nm-800 nm in 5 nm step sizes	Y	Y	Y	Y	N
2" dia. AR-coated sapphire	1	180 nm-3000 nm in 5 nm step sizes	Y*	N	N	N	Y

* It was valuable to perform transmittance and/or reflectance testing on the 2-inch diameter witness samples (one of each type) because those can go to longer wavelengths. Otherwise, one would only have relevant information out to a 800 nm wavelength.

† Proposed to test one of the six windows.

‡ Six Corning 7980 FS (uncoated) windows of 17 ± 0.1 in diameter and 1.30 ± 0.05 in thickness and one c-plane sapphire window (both sides anti reflective-coated window of 17 ± 0.1 in diameter and 0.75 ± 0.05 in thickness. The witness samples were all 2-inches in diameter, with their thickness being the same as the flight windows. The average weight of the flight windows were 10.6 kilograms and 11.2 kilograms for fused silica and AR-coated sapphire ones, respectively.

Testing Instruments and Procedures: The instruments utilized in the KSC metrology lab are summarized in Table 2. The 17-inch diameter windowpanes were secured into a test fixture and protected around the edges using microfiber cloths that were certified for contact with spacecraft windows (Figure 10). All the windowpanes came from the manufacturer with an arrow drawn on the edge of the glass, as specified on the window mechanical drawings. The windowpanes were oriented in the same position when placed in the test fixture, with the arrows pointing towards the interferometer and in the 12 o'clock (vertical) position. The panes were tested for transmittance, reflectance, color balance, wavefront, and wedge. Birefringence testing was only conducted on one 17-inch fused silica windowpane—serial number 23-0000112-001-9015.

Table 2. KSC metrology lab instrument inventory

Measurement	Instrument	Model	Wavelength	Note
Transmittance	Visible Spectrometer	N/A	320–800 nm (5 nm/step)	*
Transmittance	NIR-VIS-UV Spectrometer	PerkinElmer 1050+ spectrometer	180–3000 nm (5 nm/step)	†
Reflectance	Derived from transmission data (450 nm – 800 nm)			
Color balance	Derived from transmission data (380 nm-780 nm)			
Birefringence		N/A	N/A	‡
Transmitted Wavefront Error	Interferometer	Zygo Interferometer 6/27/24 M94937	633 nm	§
Wedge	Derived from normal Zygo data Zernike sphere terms			
Haze	NIR-VIS-UV Spectrometer	PerkinElmer	380–780 nm	†

* Newport Oriel Cornerstone 260 1/4m Monochromater along with a halogen bulb, a chopper wheel, optics, and a photodetector custom built by Dr. Mark Nurge and Dr. Bob Youngquist.

† Baseline measurements performed before tests and NIR-VIS-UV spectrometer for the Haze measurement was calibrated using the Transparent Plastic Haze Standards from HunterLab [13], [14].

‡ Custom built by Dr. Nurge and Dr. Youngquist.

§ Calibrated on June 27, 2023.

All six fused silica witness samples were tested for haze. One sample, SN01, was tested for luminous transmittance, color balance, reflectance, and haze. Due to the sample thickness (1.3-inch), the lens holder in the test chamber could not secure the sample for luminous transmittance and haze testing (Figure 11). For the luminous transmittance testing the samples were placed on a riser block covered with lens paper to protect the sample and elevate it to the correct aperture height. For haze testing, the sample was required to contact the aperture. To safely accommodate the sample in this position, it was necessary to apply Flashbreaker® tape to the sample's edge and the test chamber housing to ensure the safety of the sample. Flashbreaker® tape is certified to contact spacecraft flight windows and does not leave residue on the sample surface.

The 17-inch FS windowpanes were tested for TWE using a Zygo interferometer. The interferometer uses a 6-inch diameter aperture to take measurements on the surface of the windows to create a topographical map of the wavefront PV variation. The measurements were performed in a serpentine pattern with at least a 3 -inch overlap between shots as shown in Figure 12. The testing was performed on the windowpane in three different positions. In the first test, the windowpane surface was normal to the aperture of the interferometer. In this configuration, a total of thirty-one images were captured (Figure 12a). Wedge measurements were derived from data collected in this configuration. The following configuration involved tilting the testing fixture to 30 degrees (Figure 13). In this configuration, 20 images were captured (Figure 12b). The windowpane was then rotated 90 degrees to the right, placing the arrow drawn on the edge of the windowpane at 3 o'clock. An additional 20 images were captured in this position (Figure 12c).

The 17-inch FS windowpanes were tested for luminous transmittance using a visible light spectrometer (380–800 nm) in 5 nm steps. Color balance and reflectance measurements were derived from the transmittance data. Birefringence measurements were taken using two light sources and two photodetectors set at specific angles. The luminous transmittance and birefringence measurements were taken with the testing fixture in the normal (90 degree) position. All 2-inch diameter fused silica witness samples were tested using the PerkinElmer 1050+ Spectrophotometer. The tests for luminous transmittance, color balance, and reflectance were performed using the NIR-VIS-UV spectrometer from the wavelengths of 180 nm to 3000 nm in 5 nm step sizes. The haze test was performed using an integrating sphere [10], [13].

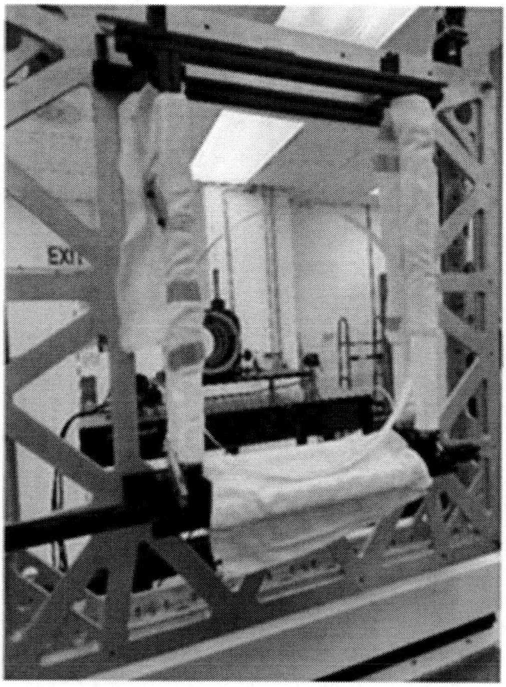

Figure 10. 17-inch FS windowpane mounted in test frame in normal (90-degree) position.

Figure 11. 2-inch FS witness sample shown during haze testing.

Only one 17-inch diameter sample (Pane Serial #: 23000112-001-9015) was proposed for birefringence measurements, by assuming that the witness samples were cut from the same billet as the 17-inch diameter windows. Birefringence is normally not an issue with glass and c-plane sapphire materials unless there is significant residual stress during the cutting. Similarly, haze testing was proposed to be performed only on witness samples, assuming the samples were of the same lot as the 17-inch diameter windows. For transmitted wavefront error testing, duplicate testing was not conducted on witness samples. Reflectance and color balance data were obtained from the measured transmittance spectra. For the color balance, it was decided to use the 1931 CIE Chromaticity Diagram [15] for the data processing[1]. Striae testing was not proposed for the windows.

The most useful information was measured using (a) for the aircraft windows and (b) for the witness samples. Overlapping data (380–800 nm visible wavelength) where possible allowed a level of confidence in using the 2-inch diameter witness sample results in the non-overlap region. For the AR-coated sapphire window, the vendor provided a window specification covering percent reflectance, not transmission, as the transmission is dependent on the substrate material property. Therefore, the reflectivity data was used to evaluate how well the AR-coating met the specifications and could be compared to data provided by the vendor. Moreover, the transmission values were incorporated into the instrument model to determine the instrument reference transmittance for the spectrophotometer and for validation of the calibration. Calibration data with the window included in the optical path remained the baseline for further analysis of the data.

Data Reduction: Data reduction is described as follows.

Transmittance, Reflectance, and Color Balance – The spectral transmittance data was obtained from the test measurement and plotted using an Excel spreadsheet. The spectrum covered a wavelength range from 320 nm to 800 nm with a wavelength increment of 5 nm. Using the spectral transmittance data and the spectral reflectance, the chromaticity coordinates shift (color balance) was calculated [9-13]. The spectral reflectance data was derived using

---

[1] The numbers to quantify the color shift are called the Chromaticity coordinates (x, y). They are computed through a given procedure that has been formulated in several published standards. The Chromaticity coordinates are normalized values of other numbers called the Tristimulus values (X, Y, Z) that are derived from the spectral transmittance data of the optical pane and the spectral characteristics of the preferred white light source, *E*( ), and of the standard observer, ( ), ( ), ( ).

$$R(\lambda) = 1 - \sqrt{T(\lambda)} \tag{1}$$

where R is the reflectance and T is the transmittance in percent. The spectral reflectance computation was calculated for each material from the transmittance data and provided with the transmittance data in an Excel spreadsheet.

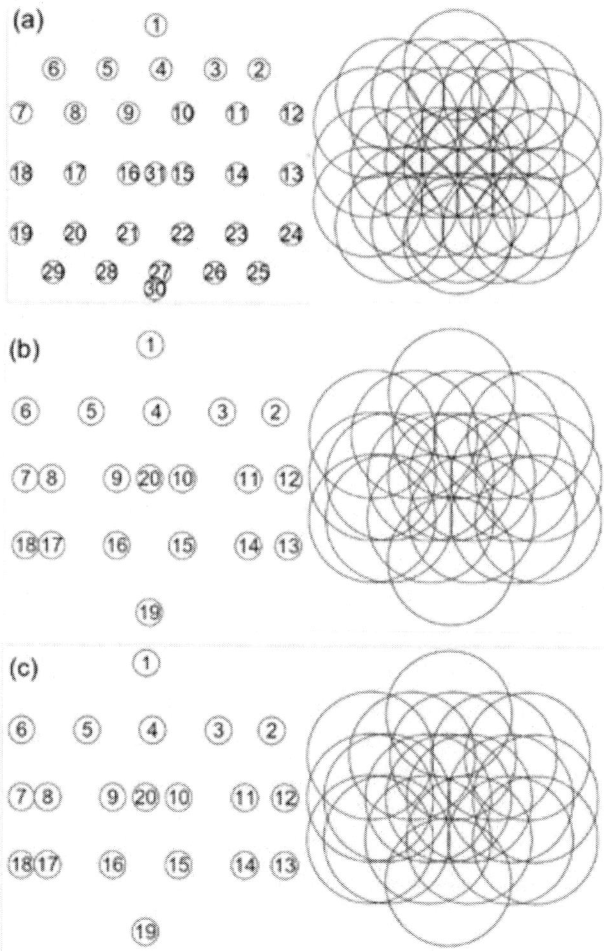

Figure 12. Scan plots and coverage maps of the interferometer uses a 6-inch diameter aperture to take measurements on the entire 17-inch diameter surface of the fused silica windows. (a) normal to the aperture position, (b) off-axis, and (c) 90 degree off-axis.

The numbers to quantify the color shift are called the chromaticity coordinates (x, y). They were computed through a procedure that has been formulated in several standards. The chromaticity coordinates are normalized values of other numbers called the tristimulus values (X, Y, Z) that were derived from the spectral transmittance data of the optical pane, the spectral characteristicsof the preferred white light source, *E* (λ), and the characteristics of the standard observer, (λ), (λ), (λ). An excel spreadsheet was created with the following data: (1) spectral transmittance of the optical pane, *T*(λ); (2) relative spectral power distribution of the illuminant, *E*(λ), and the color matching functions of the observer.

The tristimulus values X, Y, Z are the amounts of three reference color stimuli, in a given trichromatic system, required to match the color of the stimulus considered, and defined as follows:

$$X = k \int_{\lambda} E(\lambda)\, T(\lambda)\, \bar{x}(\lambda)\, d\lambda \tag{2}$$

$$Y = k \int_{\lambda} E(\lambda)\, T(\lambda)\, \bar{y}(\lambda)\, d\lambda \tag{3}$$

$$Z = k \int_{\lambda} E(\lambda)\, T(\lambda)\, \bar{z}(\lambda)\, d\lambda \tag{4}$$

where E(λ) is the relative spectral power distribution of a D65 day-light illuminant, x(λ), y(λ), and z(λ) are the color matching functions for the CIE 1931 or 1964 standard observers, T(λ)∫ is the spectral transmittance of the optical window, and k is the normalized factor $k = 100 / \int_{\lambda} E(\lambda)\, \bar{y}(\lambda)\, d\lambda$

For the 1964 standard observers, the color matching functions are often denoted by *x*10(λ), *y*10(λ), *z*10(λ), and the related tristimulus values are represented as X_{10}, Y_{10}, and Z_{10}. Except for the spectral transmittance data that is extracted from the spectrophotometer measurement, all the other quantities are tabulated in the American Society for Testing and Materials (ASTM) and International Organization for Standardization (ISO) standards [9-13]. The chromaticity coordinates x, y, and z are defined by the following equations:

$$x = \frac{X}{X+Y+Z} \tag{5}$$

$$y = \frac{Y}{X+Y+Z} \tag{6}$$

$$z = \frac{Z}{X+Y+Z} \tag{7}$$

and the chromaticity coordinates (x, y) describe how a color shift is introduced from a standard illuminant by the optical material, as shown in Figure 14.

Transmitted Wavefront Error – The raw data obtained from the wavefront measurement were contained in a binary data file that was read and processed using a Zygo software package called MetroPro® [19].

Optical Wedge – Both the wedge magnitude and wedge angle/orientation were obtained from the Zygo interferometer MetroPro® data reduction system.

Birefringence – The birefringence data reduction was automated via the acquisition system and performed point by point measurements resulting in a sequence of optical retardation points plotted as a color contour map.

Optical Haze – A version of the approach described in ASTM D1003 was utilized for the haze measurements. Table 3 presents the equations that were used to extract optical haze, with sample values obtained from FS witness sample 01 shown.

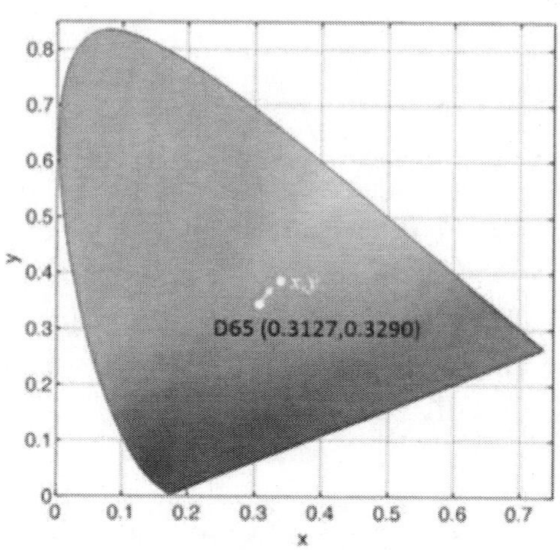

Figure 14. Color shift (x,y) from the D65 illuminant color coordinates on the 1931 standard.

Table 3. ASTM D1003 haze standard equations and sample values for FS witness sample SN01

Sample ID	ASTM D1003 Haze Standard	Value	Note
SN01	Haze =[(T4/T2) - (T3/T1)] * 100 (%)	0.01734	
	T1- Incident Light (No Sample, white plate in reflectance port)	1	Baseline
	T2- Total Light Transmitted by Specimen (Sample in transmittance position, white plate in reflectance port (total luminous transmittance)	0.9318	Y Sample and Instrument Scatter
	T3- Light Scattered by Instrument (No sample, light trap in reflectance port)	0.0141	K Part 1
	T4- Light Scattered by Instrument and Specimen (Sample in position, light trap in reflectance port (sample and instrument scatter))	0.0133	Y Instrument Scatter

Vendor Performance Data

The window vendors were required to provide NASA performance data that included as a minimum (1) surface inspection reports with defect maps, (2) inspection summary reports, (3) interferograms documenting the transmitted wavefront error (TWE), (4) interferograms documenting optical wedge (parallelism), (5) certificates of compliance with the specifications, and (6) certificates of conformance. These documents were provided by the vendor for the fused silica windows and are shown both in this section and in the appendices.

It is noted that the NASA-requested specifications were addressed but the TWE did not meet the requirements for transmitted wavefront specification. The actual measurements were as high as 0.174 waves PV. Given that a lower homogeneity grade of fused silica, the surfaces needed significant manipulation to adjust the TWE including inducing convex curves. Optical homogeneity refers to the degree of refractive index variation along the optical path of the optics. Variations in refractive index can distort the wavefront of light transmitted through the optical component and the fused silica's homogeneity is affected by bubbles and inclusions as shown in Figure 15. In the future, material of better homogeneity (λ/10 or less) would help the vendor and NASA meet the TWE specification that was quoted.

Table 4. TWE and optical wedge performance data for 17-inch FS windows

			Actual Measurements						
Characteristic	Dimension or Specification	Tolerance	*9010*	*9011*	*9012*	*9013*	*9014*	*9015*	*9016*
OD	17.000	+/-0.040"	16.981	16.981	16.980	16.982	16.982	16.981	16.990
CT	1.300	+/-0.005"	1.302	1.302	1.302	1.303	1.302	1.302	1.304
TWE	<= 0.1 waves PV		0.117	0.120	0.118	0.127	0.175	0.114	0.073
Parallelism	<=0.002"		0.001	0.000	0.001	0.000	0.002	0.000	0.000366
Bevels Side A	0.04" to 0.06" FW @ 45 deg		0.050	0.048	0.050	0.050	0.050	0.050	0.050
Bevels Side B	0.04" to 0.06" FW @ 45 deg		0.049	0.048	0.049	0.050	0.050	0.050	0.045

Table 5. TWE and optical wedge performance data for 2-inch FS witness samples

			Actual Measurements					
Characteristic	Dimension or Specification	Tolerance	*SN01*	*SN03*	*SN04*	*SN05*	*SN06*	*SN07*
OD	17.000	+/-0.040"	1.993	1.994	1.994	1.993	1.994	1.994
CT	1.300	+/-0.005"	1.3025	1.3022	1.3020	1.3022	1.3022	1.3022
TWE	<= 0.1 waves PV		0.082	0.071	0.077	0.087	0.078	0.096
Parallelism	<=0.002"		0.0000	0.0003	0.0005	0.0002	0.0005	0.0004

The fused silica window surface inspection report (S/N 9010) and window only figures (S/N 9011 – S/N 9016) from the vendor with defects identified are reproduced in Figures 15 and 16, respectively. Tables 4 and 5 summarize the TWE and optical wedge measurements conducted by the vendor along with the requirements and tolerances for the various measurands. With the exception of the TWE out of tolerance condition mentioned previously, all other performance data for the 17-inch FS windows were acceptable. There were no out of tolerances noted for the 2-inch FS witness samples. All of the windows and witness samples were considered cleared for further NASA in-house testing based on the vendor performance data.

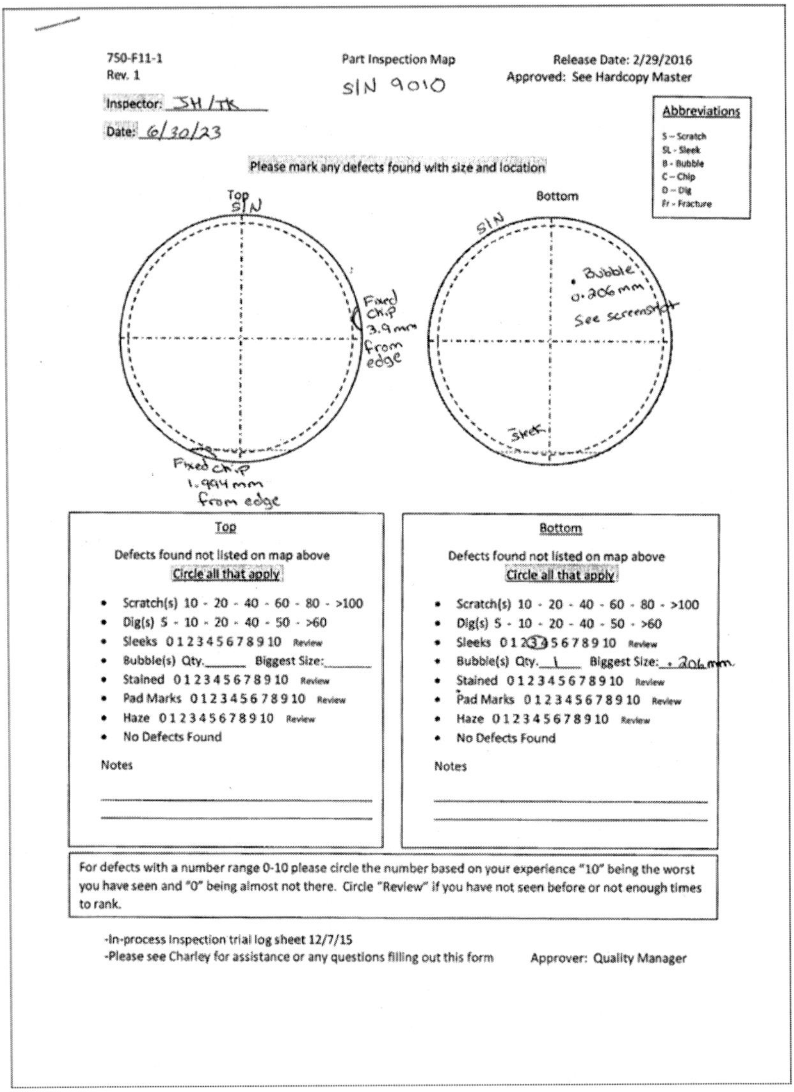

750-F11-1
Rev. 1

Part Inspection Map

S/N 9010

Release Date: 2/29/2016
Approved: See Hardcopy Master

Inspector: SH/TK

Date: 6/30/23

Abbreviations
S – Scratch
SL - Sleek
B - Bubble
C – Chip
D – Dig
Fr - Fracture

Please mark any defects found with size and location

Top S/N

Fixed chip 3.9 mm from edge

Fixed chip 1.994 mm from edge

Bottom S/N

Bubble 0.206 mm

See screenshot

Sleek

Top

Defects found not listed on map above
Circle all that apply

- Scratch(s) 10 - 20 - 40 - 60 - 80 - >100
- Dig(s) 5 - 10 - 20 - 40 - 50 - >60
- Sleeks 0 1 2 3 4 5 6 7 8 9 10 Review
- Bubble(s) Qty.______ Biggest Size:______
- Stained 0 1 2 3 4 5 6 7 8 9 10 Review
- Pad Marks 0 1 2 3 4 5 6 7 8 9 10 Review
- Haze 0 1 2 3 4 5 6 7 8 9 10 Review
- No Defects Found

Notes

Bottom

Defects found not listed on map above
Circle all that apply

- Scratch(s) 10 - 20 - 40 - 60 - 80 - >100
- Dig(s) 5 - 10 - 20 - 40 - 50 - >60
- Sleeks 0 1 2 3 4 5 6 7 8 9 10 Review
- Bubble(s) Qty. 1 Biggest Size: .206 mm
- Stained 0 1 2 3 4 5 6 7 8 9 10 Review
- Pad Marks 0 1 2 3 4 5 6 7 8 9 10 Review
- Haze 0 1 2 3 4 5 6 7 8 9 10 Review
- No Defects Found

Notes

For defects with a number range 0-10 please circle the number based on your experience "10" being the worst you have seen and "0" being almost not there. Circle "Review" if you have not seen before or not enough times to rank.

-In-process Inspection trial log sheet 12/7/15
-Please see Charley for assistance or any questions filling out this form

Approver: Quality Manager

Figure 15. Fused silica surface inspection report for window S/N 9010. The handwriting says that 'fixed chip 3.9 mm from edge' and 'fixed chip 1.994 mm from edge' on Top image and 'bubble 0.206 mm' and 'sleek' on Bottom image.

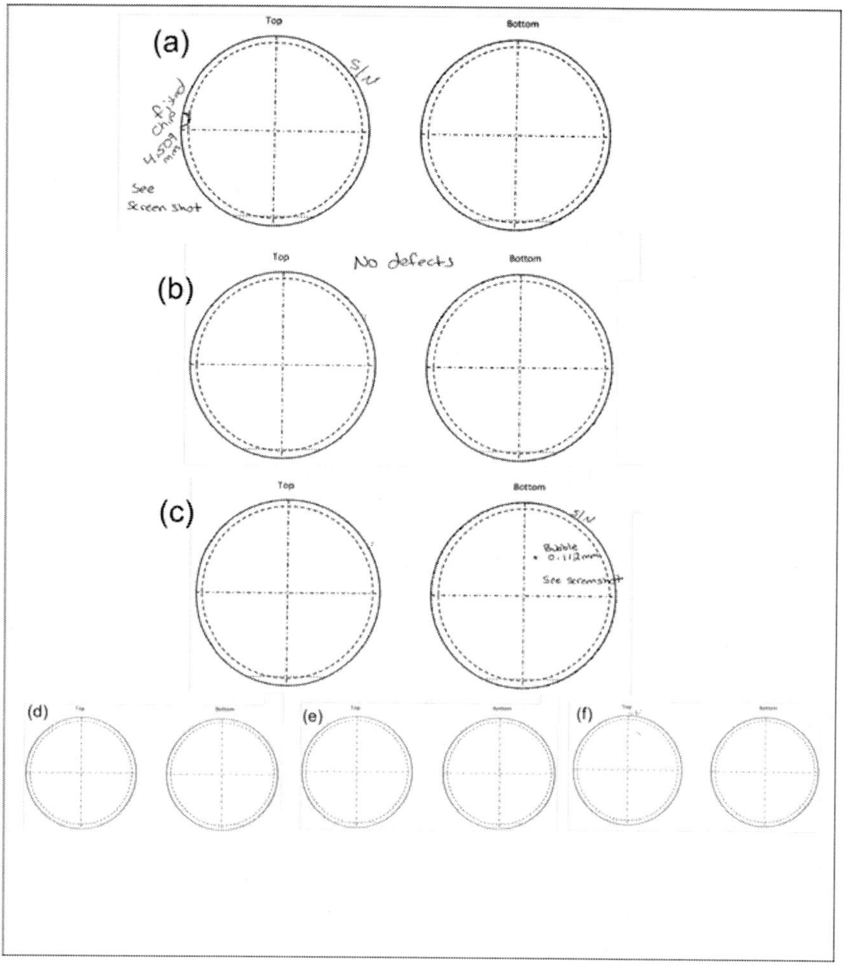

Figure 16. Fused silica surface inspection maps for window (a) S/N 9011, (b) S/N 9012, (c) S/N 9013, (d) S/N 9014, (e) S/N 9015, and (f) S/N 9016 with defects identified. The handwriting says that (a) 'fixed chip 4.509 mm' on and (c) 'bubble 0.112 mm'.

Windows Metrology Results

Figures 17–22 depict the 17-inch diameter FS windowpane testing results (included here for completeness) and consist of a summary of the data, transmittance and reflectance spectrum graphs, and a photo of the pane. One sample (pane SN 23-0000112-001-9015) contains a birefringence plot as shown in Figure 26d. In all cases, the transmittance and reflection spectra are uniform and essentially flat across the measured wavelength range of 380–800 nm. The color shift coordinates from the 1931 standard D65 illuminant color coordinates are remarkably similar between all the windows, with an average (X, Y) coordinate of (0.3130, 0.3193). The optical wedge for the windows did not exceed 0.04 arc minutes, and the TWE did not exceed 0.12 waves PV with piston and tilt removed from the measurement. Based on the vendor performance data coupled with the windows metrology results, the 17-inch FS windows were determined to be acceptable for use during the HORIS mission.

Figure 23 shows a summary of testing results for all of the 2-inch diameter FS witness samples (SN01, SN03–SN07, 1.3 in thick) including a full transmittance spectrum for one sample (SN01) and photos of a sample positioned for transmission measurements and in the test holder for haze measurements. The witness samples were the primary test articles for measurement of FS haze, and the haze measure did not exceed 0.02 for any of the samples. The transmittance spectrum depicts a fairly flat response over a wavelength range of 200–2000 nm, with a notched drop in transmission observed around 1350 nm. This drop is most likely due to the response of the light source in the photo spectrometer system.

Table 6 summarizes vendor-provided specifications for the panes along with the results of metrology testing conducted at NASA KSC. For scientific grade windows, the WFE, the amount of distortion of the wavefront through the window, should ideally be $\lambda/10$ or less. For the HORIS mission, NASA imparted a relaxed specification for WFE, better than $\lambda/8$. It is because the maintaining the requirement over the full clear aperture of the 17-inch diameter windowpanes was challenging and required the vendor to properly polish and/or coat the windows to meet the performance specification. Specifically, it was stated that for normal incidence viewing with piston, tilt, and defocus (the three most predominant aberrations) removed, the RMS wavefront error over the window would not exceed 1 wave.

Based on the results obtained from the laboratory metrology testing, it was concluded by the SCIFLI team that the FS windowpanes met the requirements for the HORIS mission.

The 17-inch diameter FS windowpane were subsequently installed on two NASA Gulfstream aircraft and used successfully in the airborne imaging campaign conducted during the OSIRIS-REx sample return capsule reentry to obtain meaningful thermal data on the capsule performance during entry and descent back to Earth in September 2023.

Table 6. Summary of windowpane vendor specifications and KSC laboratory measurements

Samples	Pane Serial # (23-0000112-001-xxxx)	Transmittance (both 17" windows and 2" witness samples)	Birefringence Testing (17" windows)	Transmitted Wavefront Error (TWE) Testing of 17" windows (vendor provided values as comparison)	Wedge Magnitude Testing of 17" windows (angle / °)	Haze Testing (2" witness samples)	Test Results
17 " dia. Fused silica (quartz)	9010	93.4 T%	Figure 32(d) map	0.107 (0.117)	0.01070 (25.02)	N	Pass
	9011	93.3		0.106 (0.120)	0.03332 (51.57)		
	9012	93.3		0.115 (0.118)	0.0278 (81.43)		
	9013	93.4		0.091 (0.127)	0.01188 (65.44)		
	9014	93.4		0.075 (0.175)	0.01671 (-18.99)		
	9015	93.4		0.074 (0.114)	0.01024 (2.11)		
2" dia. Fused silica (quartz)	SN01	93.3 T%	N	N	N	0.017%	Pass
	SN03					0.011	
	SN04					0.016	
	SN05					0.009	
	SN06					0.007	
	SN06					0.009	

Aside from surviving the flight environment, the best evidence of the windows being successful is the absence of anything of note – Signal levels and spot sizes were consistent with expectations.

As examples, data taken by SCIFLI Airborne Multispectral Imager (SAMI) payload through the FS window (S/N 9011), there were no anomalies results that have indicated a non-uniformity in the window absorption or PSF. For LaRC GIV (N552NA), the viewing angles to the SRC required looking through different parts of the windows. While coordinated translation and rotation correct for part of this, it does not fully eliminate viewing through various parts of the windows. For SAMI payload specifically, there is no vertical translation, so different portions of the window are imaged through elevation. As the view slewed through the FS windowpane, we did not see any localized variation in the sky that would indicate differences in transmission. Again, there were no anomalies in the sky or SRC spectra that required an explanation linked to pointing angles based on the results analysis by the team. Similarly, the spot formed by the SRC on the image planes did not change in size or shape as the view moved around on the window. Moreover, for the CALSPEC and NIRSPEC payloads through the FS window (S/N 9013) in JSC GII (N922NA) aircraft, fiber fed spectrometers achieving single fiber illumination from the SRC. The consistency of this result over all view angles is another example of where localized variations in the windows would have cause issues in the data. While these instruments have smaller apertures than SAMI one and therefore view through a smaller portion of the window, there was no need in the lead up to the mission to identify local "sweet spots" to image through – the optical performance was as expected with no special attention to the window required. Additional analysis could be performed to specifically look at gradients in observed sky radiance and the spot size and shape, but there have been no indications that such analysis was necessary.

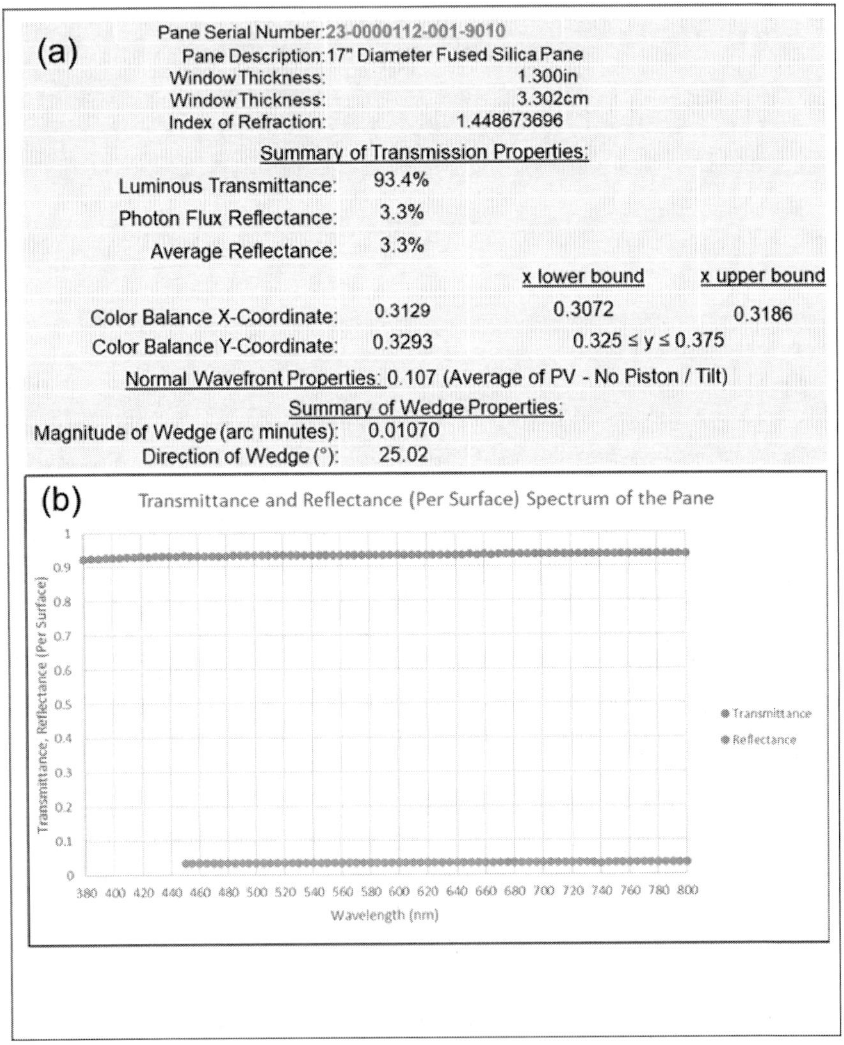

Figure 17. 17-inch FS window (SN: 23-0000112-001-9010) test results: (a) data summary, (b) transmittance / reflectance spectrum, and (c) photo of the test article.

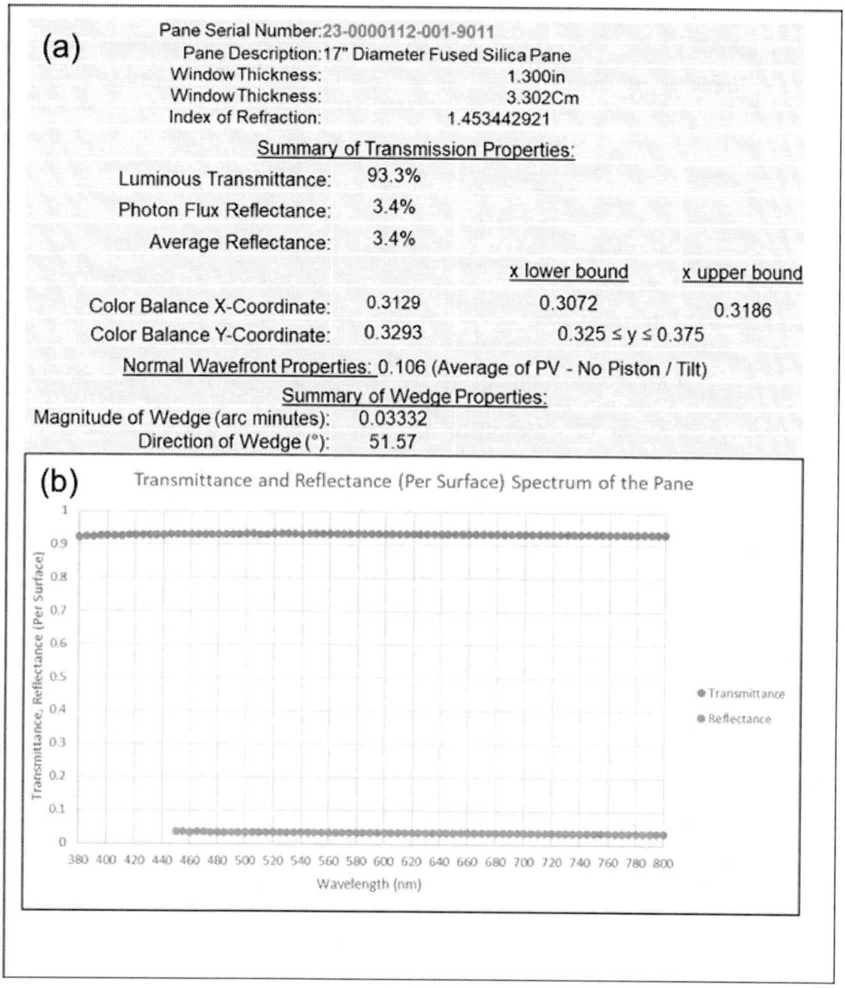
(a)

Pane Serial Number: 23-0000112-001-9011

Pane Description: 17" Diameter Fused Silica Pane

Window Thickness:	1.300in
Window Thickness:	3.302Cm
Index of Refraction:	1.453442921

Summary of Transmission Properties:

Luminous Transmittance:	93.3%		
Photon Flux Reflectance:	3.4%		
Average Reflectance:	3.4%		
		x lower bound	x upper bound
Color Balance X-Coordinate:	0.3129	0.3072	0.3186
Color Balance Y-Coordinate:	0.3293	0.325 ≤ y ≤ 0.375	

Normal Wavefront Properties: 0.106 (Average of PV - No Piston / Tilt)

Summary of Wedge Properties:

Magnitude of Wedge (arc minutes):	0.03332
Direction of Wedge (°):	51.57

Figure 18. 17-inch FS window (SN: 23-0000112-001-9011) test results.

(a)

Pane Serial Number:23-0000112-001-9012

Pane Description:17" Diameter Fused Silica Pane

Window Thickness:	1.300in
Window Thickness:	3.302cm
Index of Refraction:	1.453896032

Summary of Transmission Properties:

		x lower bound	x upper bound
Luminous Transmittance:	93.3%		
Photon Flux Reflectance:	3.4%		
Average Reflectance:	3.4%		
Color Balance X-Coordinate:	0.3130	0.3072	0.3186
Color Balance Y-Coordinate:	0.3292	$0.325 \leq y \leq 0.375$	

Normal Wavefront Properties: 0.115 (Average of PV - No Piston / Tilt)

Summary of Wedge Properties:

Magnitude of Wedge (arc minutes):	0.02787
Direction of Wedge (°):	81.43

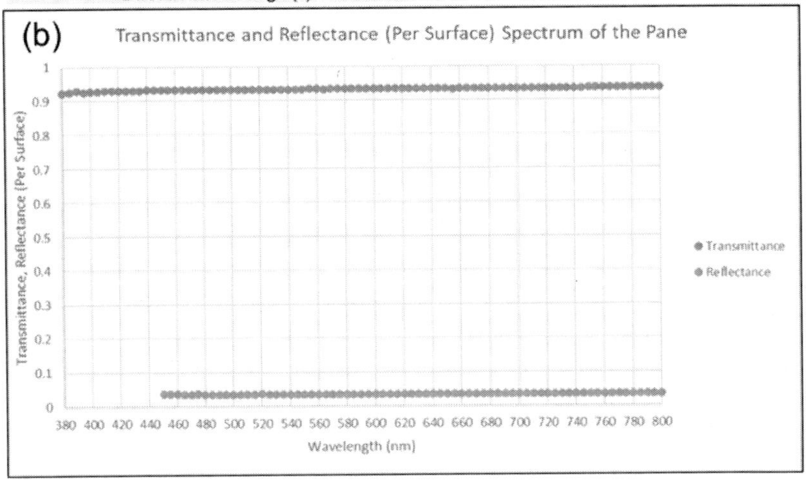

Figure 19. 17-inch FS window (SN: 23-0000112-001-9012) test results.

(a)

Pane Serial Number: 23-0000112-001-9013
Pane Description: 17" Diameter Fused Silica Pane
Window Thickness: 1.300in
Window Thickness: 3.302cm
Index of Refraction: 1.448673696

Summary of Transmission Properties:

		x lower bound	x upper bound
Luminous Transmittance:	93.4%		
Photon Flux Reflectance:	3.4%		
Average Reflectance:	3.4%		
Color Balance X-Coordinate:	0.3130	0.3072	0.3186
Color Balance Y-Coordinate:	0.3294	0.325 ≤ y ≤ 0.375	

Normal Wavefront Properties: 0.091 (Average of PV - No Piston / Tilt)

Summary of Wedge Properties:

Magnitude of Wedge (arc minutes): 0.01188
Direction of Wedge (°): 65.44

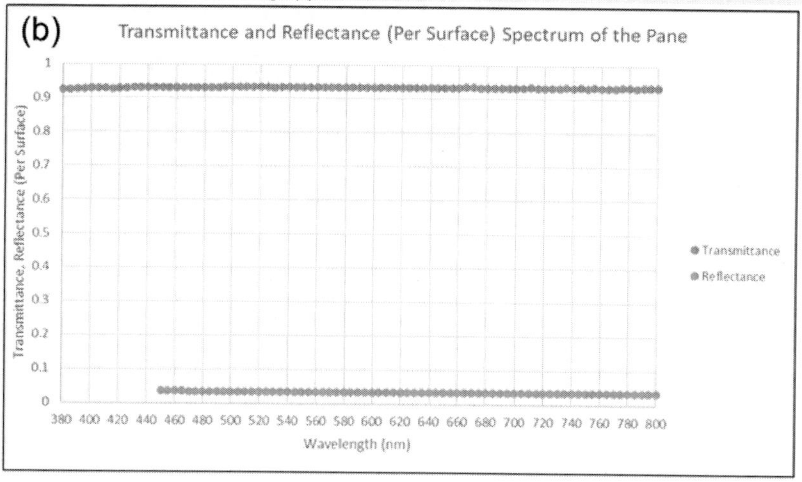

Figure 20. 17-inch FS window (SN: 23-0000112-001-9013) test results.

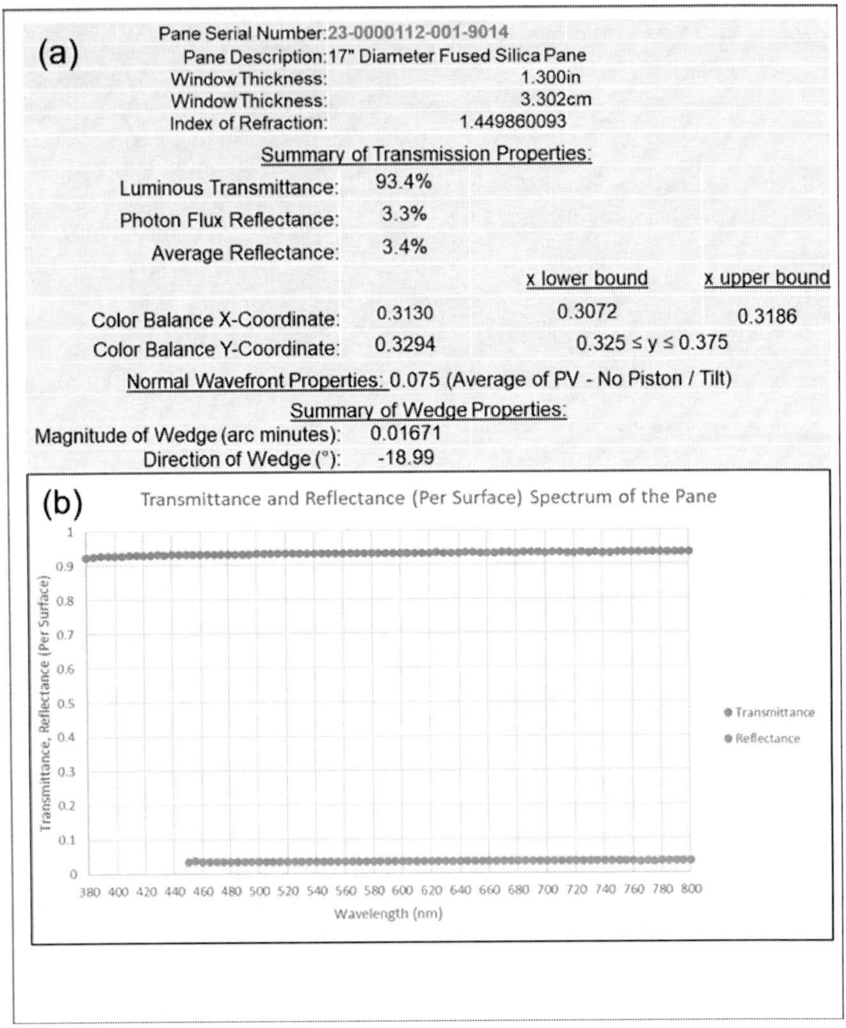

Figure 21. 17-inch FS window (SN: 23-0000112-001-9014) test results.

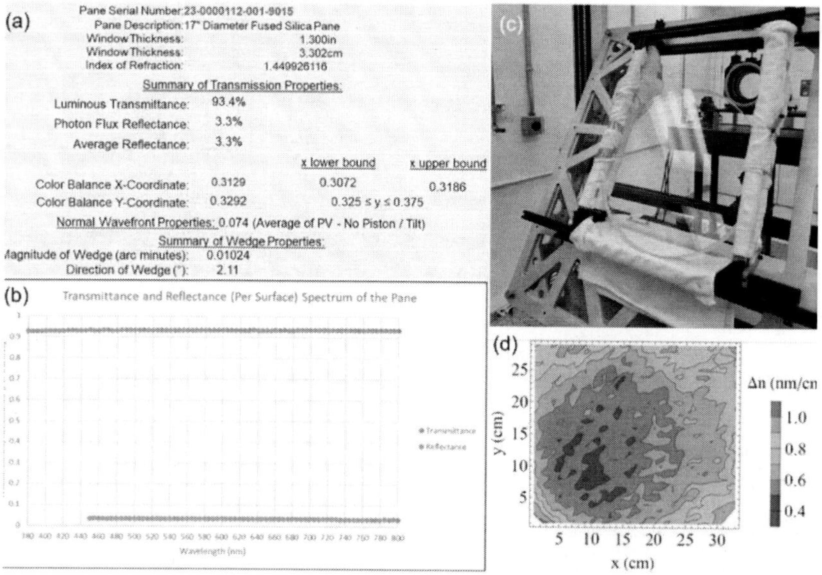

Figure 22. 17-inch FS window (SN: 23-0000112-001-9015) test results.

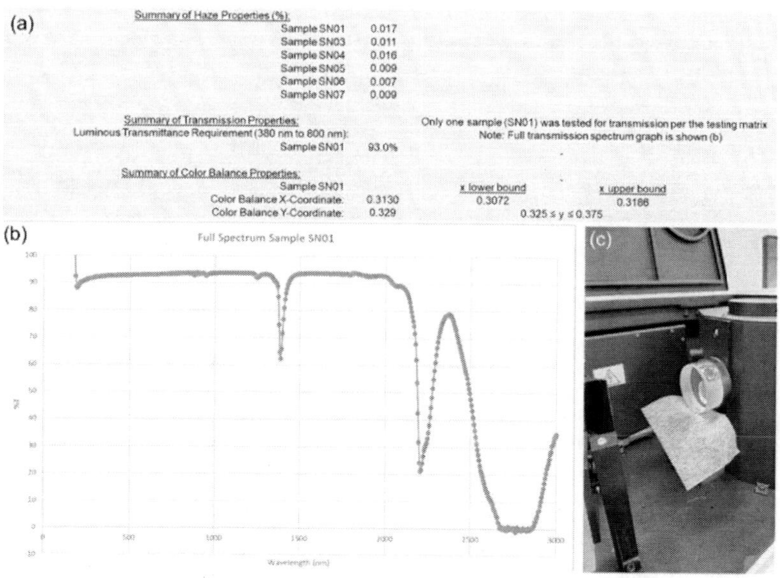

Figure 23. 2-inch diameter FS witness sample (SN01, SN03 – SN07, 1.3 in thick) test results: (a) summary of data, (b) full spectrum of SN01, (c) photo of sample positioned for transmission and haze measurements in test holder.

Summary

This report summarizes windowpane optical analyses conducted for aircraft windows designed and procured for the SCIFLI HORIS mission. Six 17-inch diameter FS windows were designed and purchased for use in airborne imaging campaigns for HORIS and subsequent. The FS panes were tested based on standard optical window characterizations including transmittance, reflectance, color balance, birefringence, transmitted wavefront error, optical wedge, and optical haze. For scientific grade windows, the WFE, the amount of distortion of the wavefront through the window, should ideally be $\lambda/10$ or less. 1/10 serves as a standard for high-quality optics. Maintaining the requirement over the full clear aperture of the windows was challenging and required the vendor to properly polish and/or coat the windows to meet the performance specification. For the HORIS mission, NASA imparted a relaxed specification for WFE, better than $\lambda/8$, depending on how the measurements were obtained and handled. Specifically, it was stated that for normal incidence viewing with piston, tilt, and defocus (the three most predominant aberrations) removed, the RMS wavefront error over the window would not exceed 1 wave.

Based on the results obtained from the laboratory metrology testing, it was concluded by the SCIFLI team that the FS windowpanes met the requirements for the HORIS mission. They were subsequently installed on two NASA Gulfstream aircraft and used successfully in the airborne imaging campaign conducted during the OSIRIS-REx sample return capsule reentry to obtain meaningful thermal data on the capsule performance during entry and descent back to Earth in September 2023.

Acknowledgments

1) Testing objectives and the data acquisition and processing sections are formulated based on work by Dr. Ken Tedjojuwono (LaRC Distinguished Research Associate), through internal NASA reports entitled: (1) Transmittance, Reflectance and Color Balance Testing in Support of the Window Material Database Test Plan, (2) Birefringence Testing in Support of the Window Material Database Test Plan, (3) TWE Testing in Support of the Window Material

Database Test Plan, and (4) Wedge Testing in Support of the Window Material Database Test Plan.

2) The testing results section is expanded based "SCIFLI OSIRIS-REx Windowpane Optical Analysis Preliminary Draft" (Document ID: NASA-5557, Rev: Basic) written by Ian Rook and Thomas Gutierrez in the Laboratories, Development and Testing Division (LDTD) in the Engineering Directorate (ED) at the John F. Kennedy Space Center (KSC).

Appendix A. Wavefront Interferograms for FS Windows

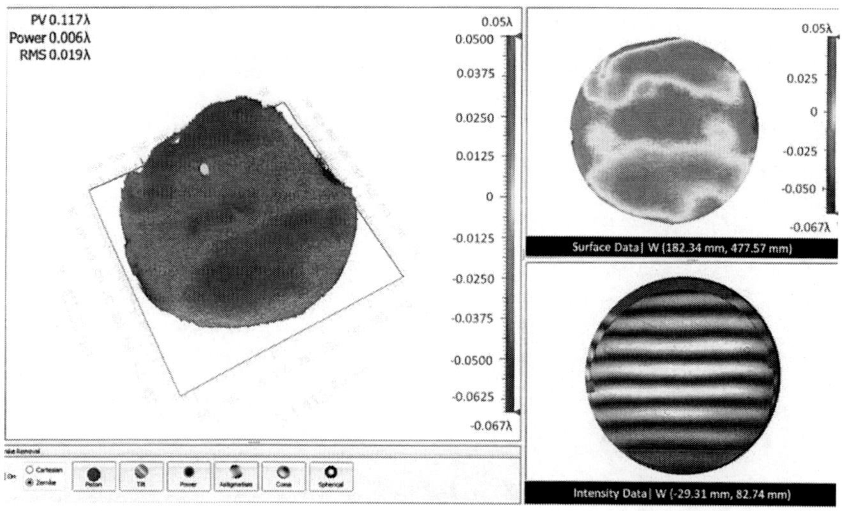

Figure A1. Interferogram for 17-inch FS window, SN 23000112-001-9010.

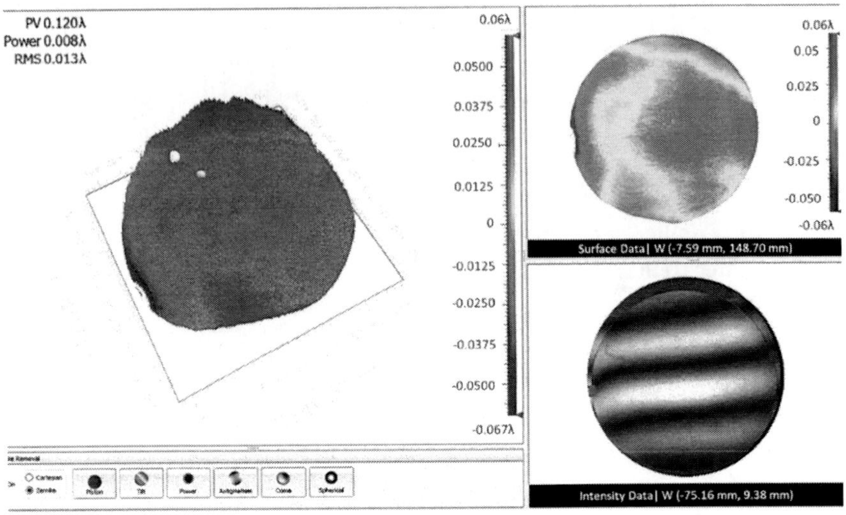

Figure A2. Interferogram for 17-inch FS window, SN 23000112-001-9011.

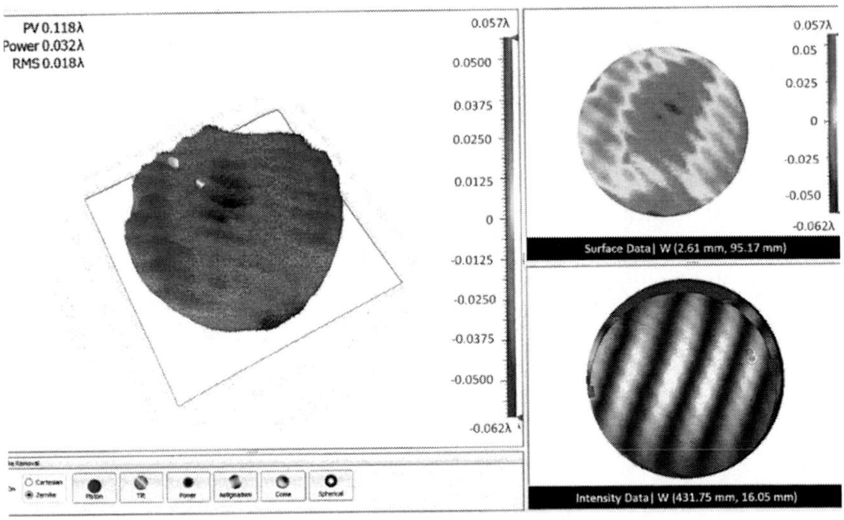

Figure A3. Interferogram for 17-inch FS window, SN 23000112-001-9012.

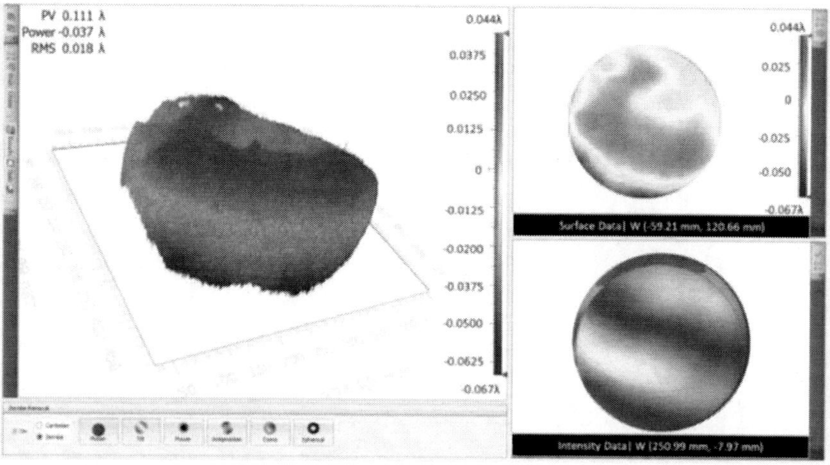

Figure A4. Interferogram for 17-inch FS window, SN 23000112-001-9013.

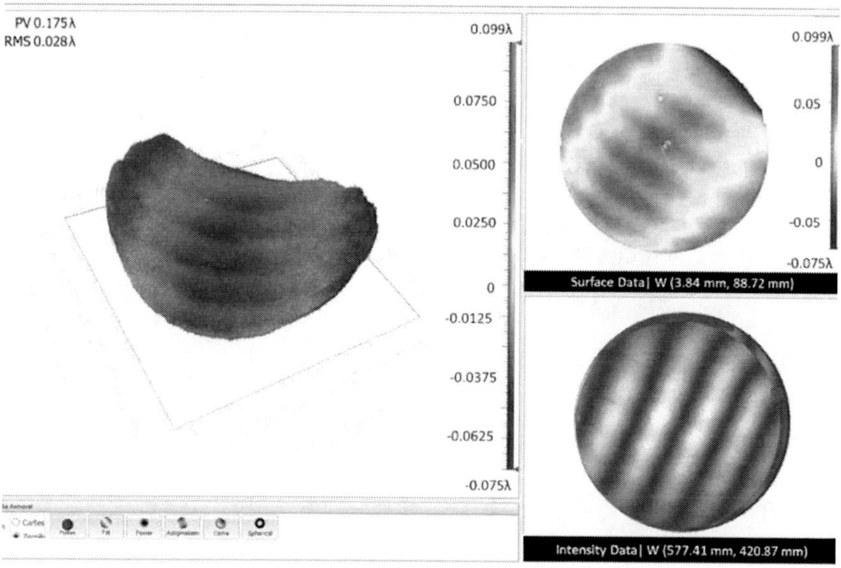

Figure A5. Interferogram for 17-inch FS window, SN 23000112-001-9014.

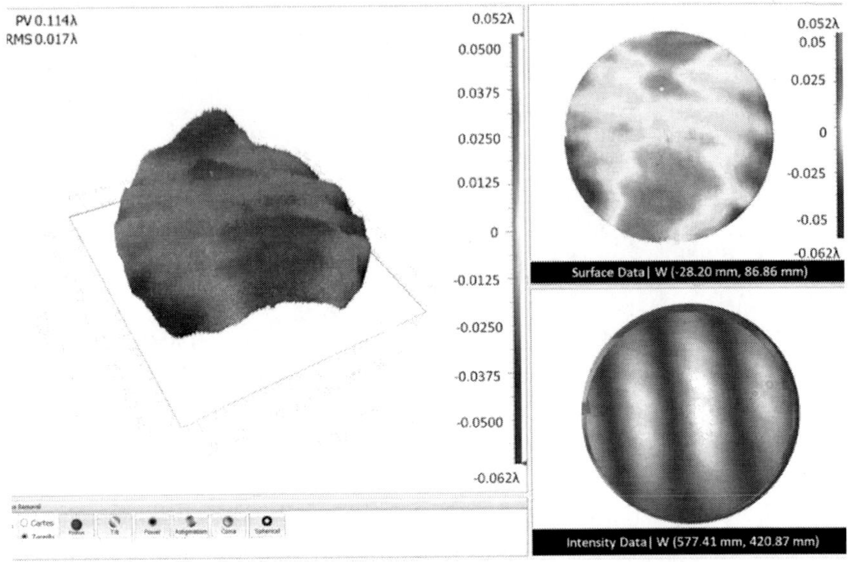

Figure A6. Interferogram for 17-inch FS window, SN 23000112-001-9015.

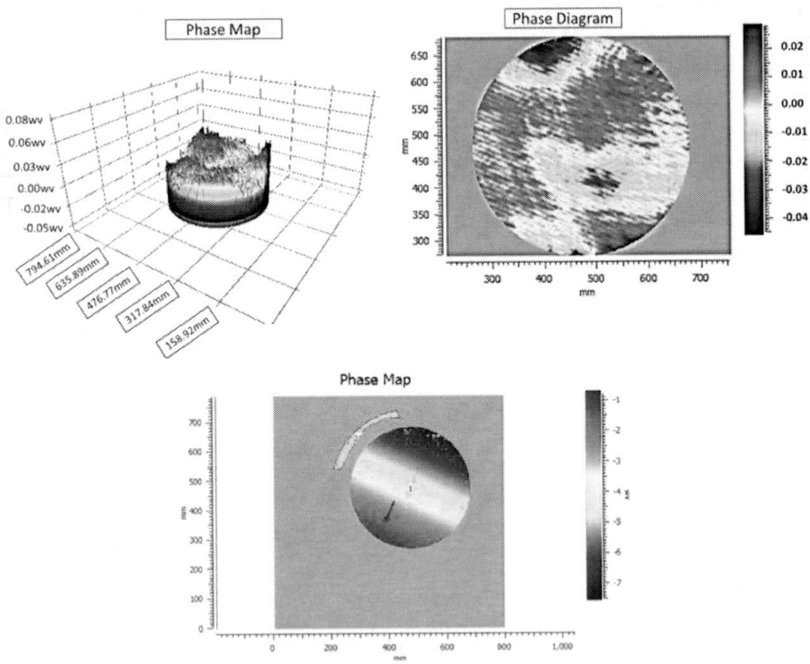

Figure A7. Interferogram result for 17-inch FS window, SN 23000112-001-9016.

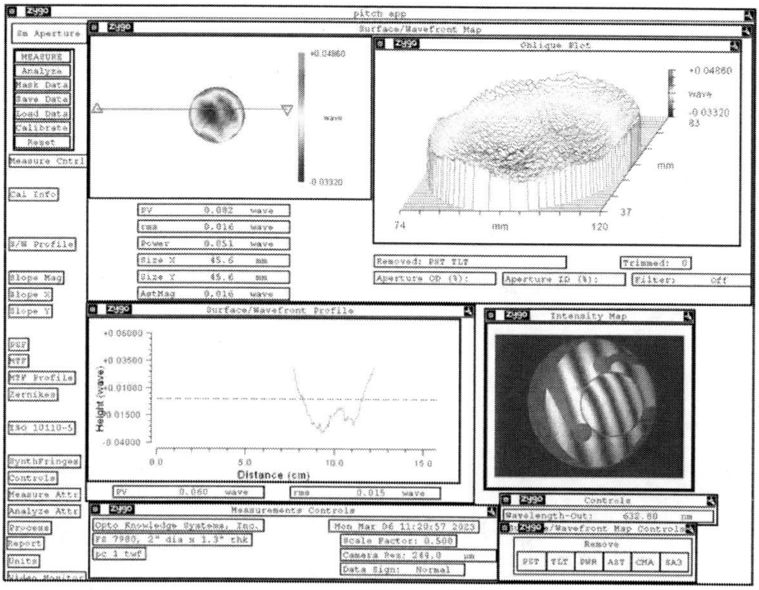

Figure A8. TWE Interferogram for 2-inch FS witness sample, SN01.

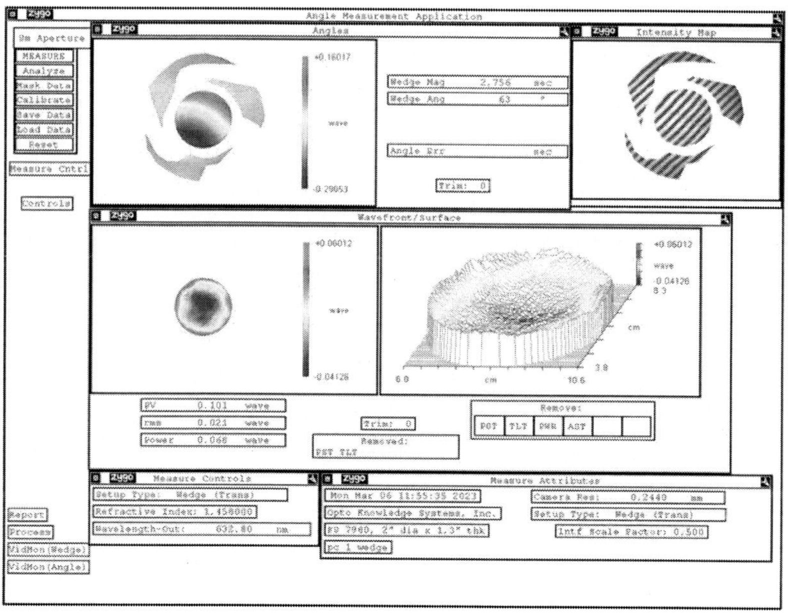

Figure A9. Wedge Interferogram for 2-inch FS witness sample, SN01.

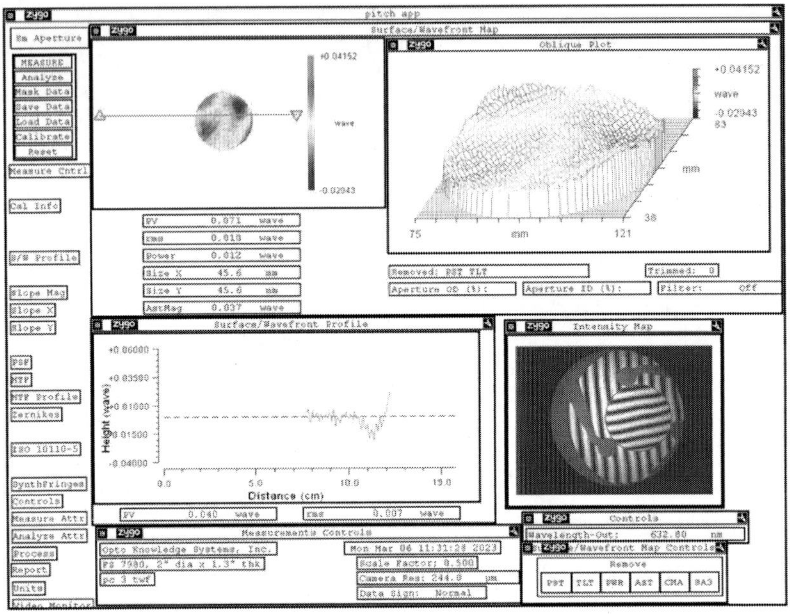

Figure A10. TWE Interferogram for 2-inch FS witness sample, SN03.

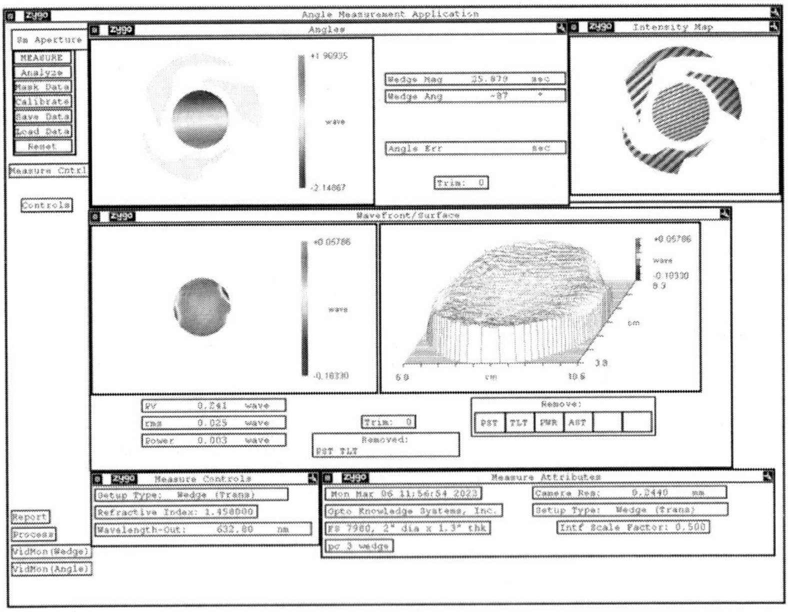

Figure A11. Wedge Interferogram for 2-inch FS witness sample, SN03.

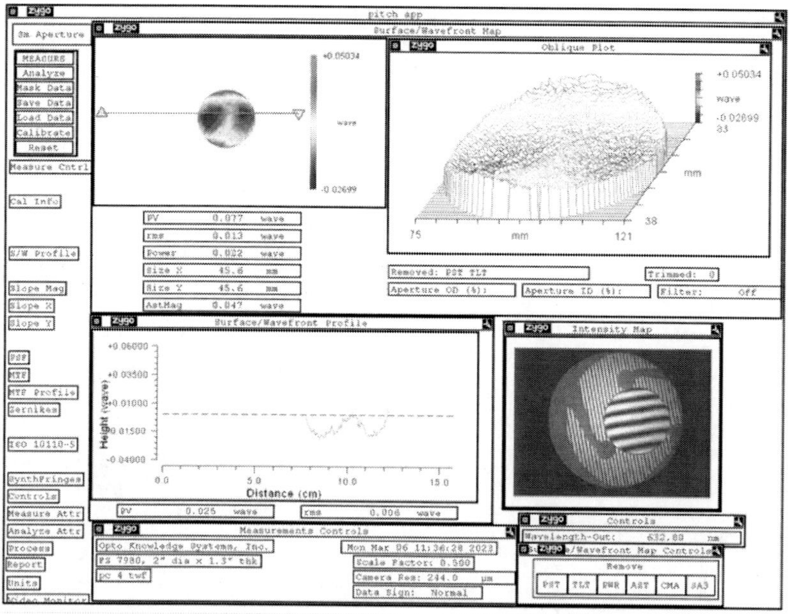

Figure A12. TWE Interferogram for 2-inch FS witness sample, SN04.

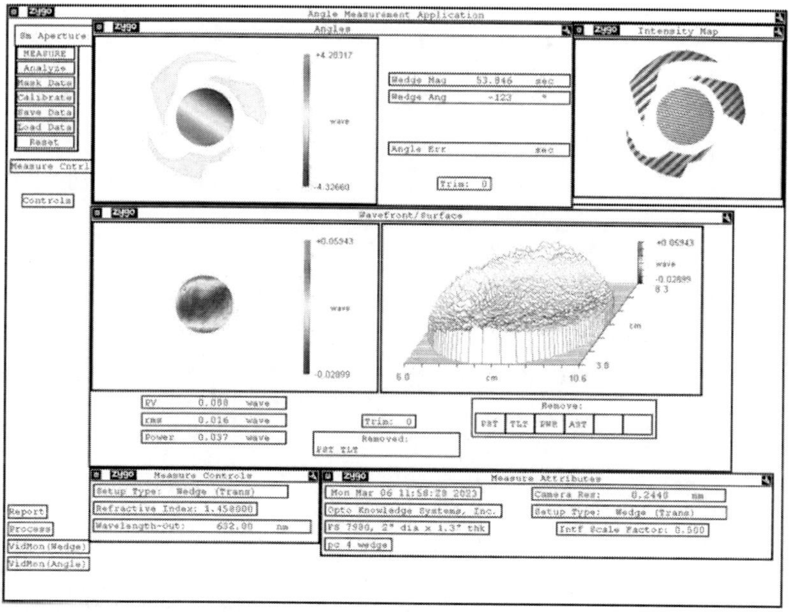

Figure A13. Wedge Interferogram for 2-inch FS witness sample, SN04.

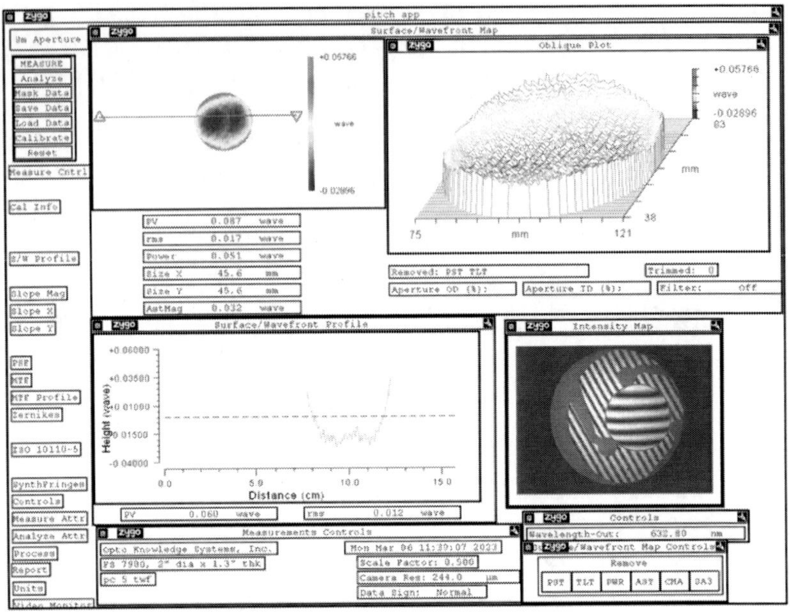

Figure A14. TWE Interferogram for 2-inch FS witness sample, SN05.

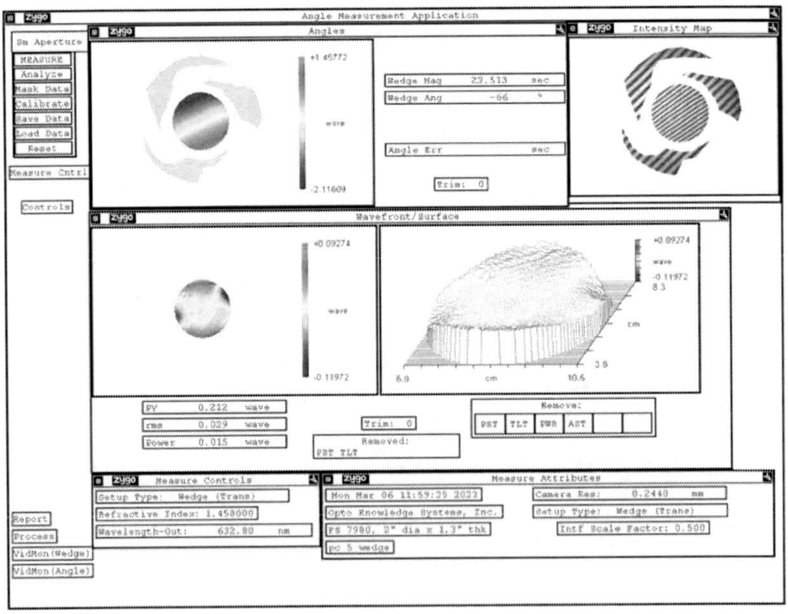

Figure A15. Wedge Interferogram for 2-inch FS witness sample, SN05.

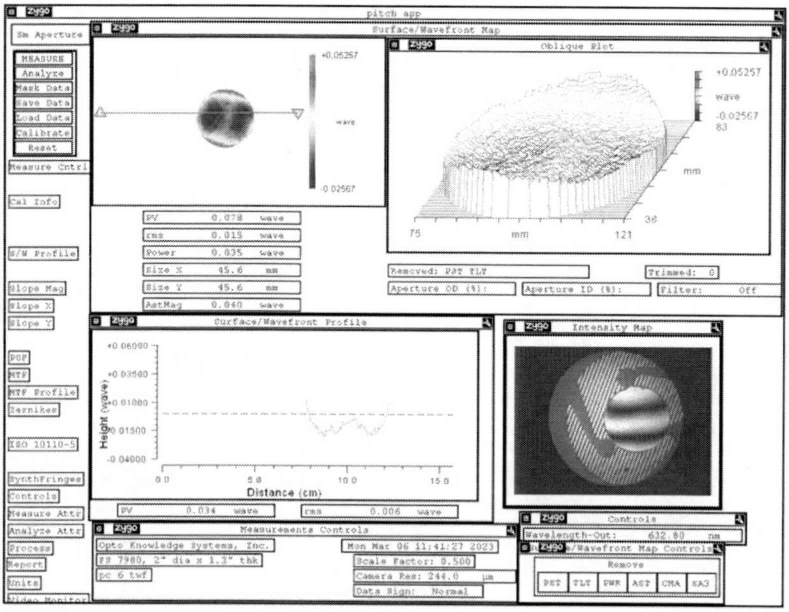

Figure A16. TWE Interferogram for 2-inch FS witness sample, SN06.

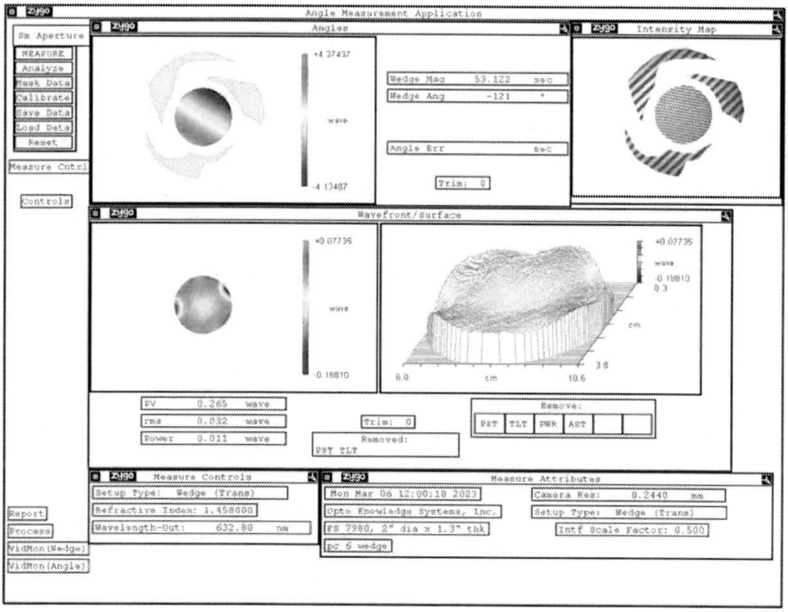

Figure A17. Wedge Interferogram for 2-inch FS witness sample, SN06.

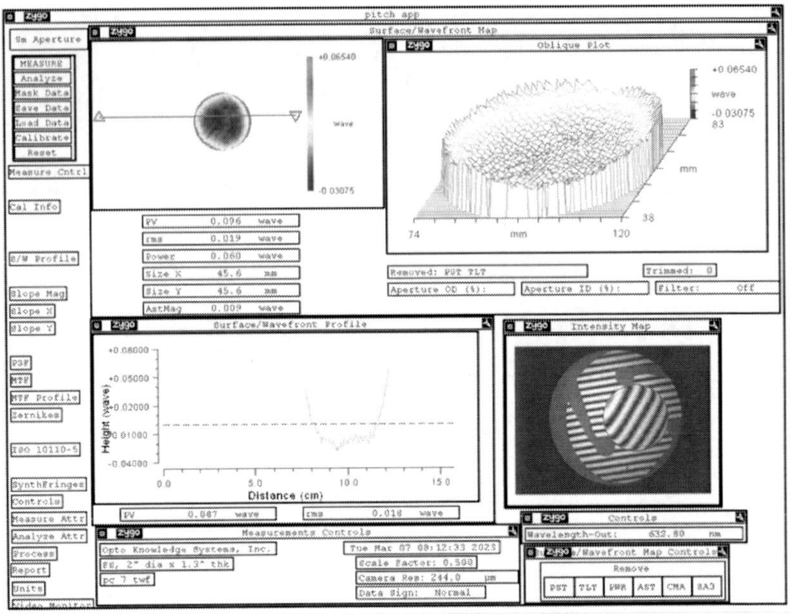

Figure A18. TWE Interferogram for 2-inch FS witness sample, SN07.

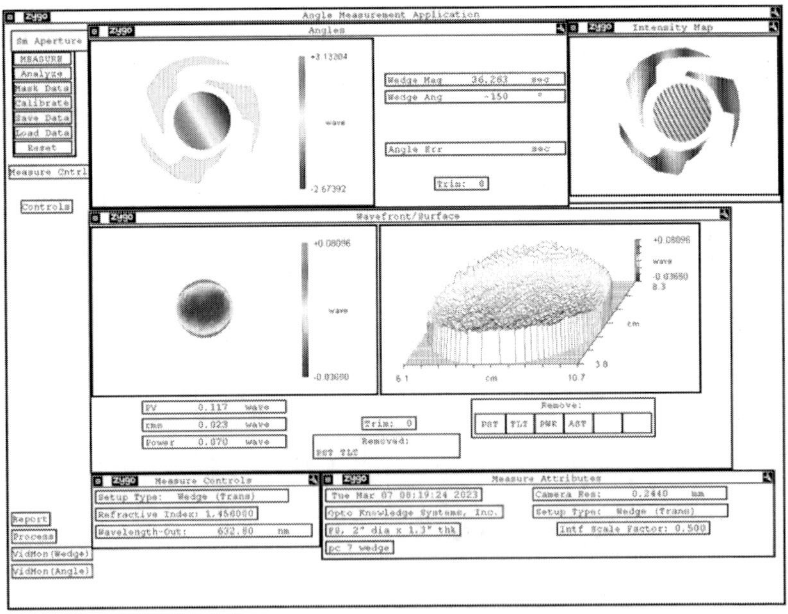

Figure A19. Wedge Interferogram for 2-inch FS witness sample, SN07.

References

[1] D. Vigna, L. Faggioli, A. Milani, F. Spoto, D. Farnocchia, and B. Carry, "Detecting the Yarkovsky effect among near-Earth asteroids from astrometric data," A&A, vol. 617, p. A61, Sep. 2018, doi: 10.1051/0004-6361/201833153.

[2] "SCIFLI – In-Flight Imagery for Today's Space Age." Accessed: Oct. 16, 2023. [Online]. Available: https://scifli.larc.nasa.gov/

[3] "U.S. Military Performance Specification. (1997). General specification governing the manufacturing, assembly, and inspection of Optical Components for Fire Control Instruments (Mil-PRF-13830B)."

[4] "OSIRIS-REx - NASA Science." Accessed: Oct. 09, 2023. [Online]. Available: https://science.nasa.gov/mission/osiris-rex.

[5] "OSIRIS-REx: Discovering the Origins of the Solar System | Lockheed Martin." Accessed: Oct. 09, 2023. [Online]. Available: https://www.lockheedmartin.com/en-us/products/osiris-rex.html.

[6] "OSIRIS-REx Mission," OSIRIS-REx Mission. Accessed: Oct. 09, 2023. [Online]. Available: https://www.asteroidmission.org/.

[7] "Apophis - NASA Science." Accessed: Feb. 16, 2024. [Online]. Available: https://science.nasa.gov/solar-system/asteroids/apophis/.

[8] "CIE 1931 color space," Wikipedia. Feb. 20, 2024. Accessed: Feb. 21, 2024. [Online]. Available: https://en.wikipedia.org/w/index.php?title=CIE_1931_color_space&oldid=1209165237.

[9] "ASTM D1925 Standard Test Method for Yellowness Index of Plastics."

[10] "ASTM E308 – 08 Computing the Colors Using the CIE System."

[11] "ASTM E313 – 10 Standard Practice for Calculating Yellowness and Whiteness Indices from Instrumentally Measured Color Coordinates."

[12] "ISO 11664-1:2007(E) Colorimetry — Part 1: CIE standard colorimetric observers."

[13] "ISO 11664-2:2007(E) Colorimetry — Part 2: CIE standard illuminants."

[14] K. K. Tedjojuwono, N. Clark, and W. M. Humphreys, "Optical characterization of window materials for aerospace applications," presented at the SPIE Optifab, J. L. Bentley and M. Pfaff, Eds., Rochester, New York, USA, Sep. 2013, p. 88841I. doi: 10.1117/12.2029469.

[15] Parris, "Practical Guide to Specifying Optical Components." Arizona State University, 2012. [Online]. Available: https://wp.optics.arizona.edu/optomech/wp-content/uploads/sites/53/2016/10/Practical-Guide-to-Specifying-Optical-Components.pdf.

[16] P. R. Y. Jr, Opto-Mechanical Systems Design. CRC Press, 2005.

[17] R. C. Youngquist, M. Nurge, S. Danley, and M. G. Skow, "Category B Plastic Pane Testing for JSC 66320 Rev A Requirements Verifiable Through Acceptance Testing," NASA/TM-2016-219156, Aug. 2016. Accessed: Mar. 01, 2024. [Online]. Available: https://ntrs.nasa.gov/citations/20180003314.

[18] "Source for ASTM D1003 Transparent Plastic Haze Standards from HunterLab," Hunterlab. Accessed: Mar. 01, 2024. [Online]. Available: https://support.hunterlab.com/hc/en-us/articles/202023275-Source-for-ASTM-D1003-Transparent-Plastic-Haze-Standards-from-HunterLab.

[19] "Metrology Software." Accessed: Mar. 03, 2024. [Online]. Available: https://www.zygo.com/en/support/metrology-software

Index

I

J

L

M

N

O

P

R

S